The Making of Thoreau's Excursions

The Making of Thoreau's Excursions

Disturbance Ecologies in Antebellum US Travel Writing

Jake McGinnis

BLOOMSBURY ACADEMIC
NEW YORK • LONDON • OXFORD • NEW DELHI • SYDNEY

BLOOMSBURY ACADEMIC
Bloomsbury Publishing Inc, 1359 Broadway, New York, NY 10018, USA
Bloomsbury Publishing Plc, 50 Bedford Square, London, WC1B 3DP, UK
Bloomsbury Publishing Ireland, 29 Earlsfort Terrace, Dublin 2, D02 AY28, Ireland

BLOOMSBURY, BLOOMSBURY ACADEMIC and the Diana logo are trademarks of Bloomsbury Publishing Plc

First published in the United States of America 2026

Cover design: Eleanor Rose
Cover image: Appalachian Trail, Mount Katahdin, USA © Evan Jarvi / Getty Images

Library of Congress Cataloging-in-Publication Data

ISBN: HB: 979-8-7651-6356-6
ePDF: 979-8-7651-6355-9
eBook: 979-8-7651-6358-0

Typeset by Newgen KnowledgeWorks Pvt. Ltd., Chennai, India

For product safety related questions contact productsafety@bloomsbury.com.

In memory of Nancy Lee Kaminski,
who first showed me how to travel differently.

Contents

Acknowledgments

Writing often feels like a solitary endeavor, practiced in quiet places amid a shroud of secrecy. In reality, though, books emerge amidst communities. Many people have shaped this project, far more than I can mention here. This project would not have been possible without Laura Dassow Walls, who enthusiastically welcomed me to the field of Thoreau studies as a young scholar of literature and environment and who assisted in every stage of this project. In our many conversations and email exchanges and in her own writing and scholarship, she has been both a guide and a steady source of inspiration. This book would simply not be possible without her support and encouragement. I am also thankful for the advice and guidance of Scott Slovic, who likewise supported this project since the beginning. I first began thinking about travel writing and environmental literature with him in a seminar at the University of Idaho, and he has been an unwavering source of support since. It was Scott who suggested, in the middle of a morning run in Davis, California, that I consider writing about *Horizon* at some point in these pages. Sandra Gustafson and Nan Da also offered invaluable support and, in the range and creativity of their scholarship, vital inspiration.

Many other people have shaped this book, both directly and in less tangible ways. I drafted many of these pages among friends at the University of Notre Dame, and I am particularly thankful for help and good cheer from Josh Wright, Joel Gabriel Kempff, Kade Ivy, Logan Quigley, Sara Judy, Claudia Carrol, Anila Shree, Julian Dean, Stacy Sivinski, Caitlin Smith, Tyler Gardner, Jay Miller, and Aleks Hernandez, all of whom listened to me talk about this project or actually sat next to me (on Zoom and sometimes even in the "actual" world) while I worked. All of these good people have inspired something within these pages, and I'm grateful for their continued friendship.

I am grateful, too, for the friendship and conversation among the ecocriticism reading group in Idaho, especially Erin James, Jennifer Ladino, Peter Remien, Xinmin Liu, Anna Banks, Debbie Lee, Donna Potts, and Kota Inoue. Those discussions offered a wonderful place to begin thinking about literature and environment. I'm also incredibly lucky to have worked with Gary Williams early in my graduate studies, for he helped me begin writing about nineteenth-century American literature in earnest, and likewise Walter Hesford, who shared my interest in Thoreau and offered me some crucial books, which are still on my shelves today. My thanks also to Alan Brew, who encouraged me to study literature in the first place and who has shared some memorable excursions near Lake Superior, including on that beach with a long view toward La Pointe, and at the place Copway called "Burnt Wood River."

The Thoreau Society, now the Thoreau Alliance, provided an essential and exceptionally welcoming environment for a young scholar. In particular, I'd like to single out Rochelle Johnson, John Kucich, James Finley, Henrik Otterberg, Mike Berger, and Ron Hoag for good conversation, advice, and time on the water and on the trail. Many friends from the Association for the Study of Literature and Environment deserve my gratitude as well, far too many for me to mention here. I'd like to thank Jim Warren, Lowell Duckert, and Catherine Bowlin, who read early drafts of these chapters and offered both encouragement and very helpful suggestions, as well as Andrew Rose, Eric Morel, Cory Willard, and Emily Rau for asking some helpful and memorable questions.

My colleagues at Graceland University provided crucial support toward completing this book, and I also received support from the Graduate School, the College of Arts and Letters, and the Graduate Student Government at the University of Notre Dame. Portions of my work here were supported by a Thoreau Society Graduate Student Fellowship, which offered a wonderful opportunity to conduct research in Concord. I'm also grateful to the Notre Dame Institute for Advanced Studies, now the Institute for Ethics and the Common Good, which offered both additional research funding and an exceptional environment for testing out new ideas, helping me to see the real value of interdisciplinary scholarship and collaboration.

My thanks to Amy Martin, Hali Han, Jaxson Mannes, and the rest of the staff at Bloomsbury for their attention and care with this manuscript. Also, I'm grateful for the reviewers, who offered actionable, encouraging feedback on this project. Thank you for demonstrating that peer review can and does, in the best of cases, support a writer's work. Portions of Chapter 4 were originally published in *Studies in American Indian Literature*, and I am grateful to the editors and reviewers for their feedback and assistance, and to the University of Nebraska Press for permission to reprint a revised version of that essay here.

As a writer and a teacher-scholar who focuses on travel and place, I am also grateful to the landscapes and communities in which I have lived and traveled while completing this book. In particular, I am fortunate to have visited Katahdin and the Penobscot watershed and to have explored the many protected lands around Concord, the same swamps and woodlands where Thoreau practiced his daily excursions. I'm also grateful for my time in the Lake Michigan basin and now the Tallgrass Prairie, places that have taught me a great deal about the environments described by the authors I study.

Finally, and above all, I am grateful to Celeste, for everything, including the many travels that we've shared, and to Mel, who shows me every day how to move through the world with grace and joy.

Abbreviations

CC Henry David Thoreau, *Cape Cod. The Writings of Henry D. Thoreau.* Edited by Joseph J. Moldenhauer. Princeton, NJ: Princeton University Press, 1988.

Corr Henry David Thoreau, *The Correspondence of Henry David Thoreau.* Edited by Robert N. Hudspeth, Elizabeth Hall Witherell, and Lihong Xie. 2 vols. Princeton, NJ: Princeton University Press, 2013. [Volumes indicated by Arabic numerals.]

Exc Henry David Thoreau, *Excursions. The Writings of Henry D. Thoreau.* Edited by Joseph J. Moldenhauer. Princeton, NJ: Princeton University Press, 2007.

Life George Copway, *Life, Letters, and Speeches of George Copway (Kahgegagahbowh).* Edited by A. LaVonne Brown and Donald B. Smith. Lincoln: University of Nebraska Press, 1997.

Lot George William Curtis, *Lotus-Eating: A Summer Book.* New York: Harper & Brothers, 1852.

MW Henry David Thoreau, *The Maine Woods. The Writings of Henry D. Thoreau.* Edited by Joseph J. Moldenhauer. Princeton, NJ: Princeton University Press, 1972.

PJ Henry David Thoreau, *Journal. The Writings of Henry D. Thoreau.* Edited by Elizabeth Hall Witherell, William L. Howarth, Robert Sattelmeyer, Thomas Blanding, Mark R. Patterson, William Rossi, Leonard N. Neufeldt, Nancy Craig Simmons, Patrick F. O'Connell, Heather Kirk Thomas, Ron Thomas, and Sandra Harbert Petrulionis, 8 vols. Princeton, NJ: Princeton University Press, 1981–.

Summer Margaret Fuller, *Summer on the Lakes, in 1843.* Champaign: University of Illinois Press, 1991.

Tour Washington Irving, *A Tour on the Prairies.* In *Three Western Narratives*, edited by James P. Ronda, 1–162. New York: Library of America, 2004.

Wa Henry David Thoreau, *Walden.* Edited by J. Lyndon Shanley. Princeton, NJ: Princeton University Press, 1971.

Wk Henry David Thoreau, *A Week on the Concord and Merrimack Rivers.* Edited J. Lyndon Shanley. Princeton, NJ: Princeton University Press, 1971.

Introduction: The Excursion

In an August 18, 1857 letter, Henry David Thoreau described his longest and most ambitious excursion to the Maine Woods, a 200-mile, fifteen-day route from Moosehead Lake through the infamous Mud Pond Carry to the East Branch of the Penobscot River. Over the course of the journey, he'd been hopelessly lost; he'd argued with his guide and panicked over the fate of his traveling companion, Edward Hoar; and he'd had to abandon plans to climb Katahdin, the mountain that he'd dreamed of for years. Still, the trip was exhilarating, "chiefly from associating with an intelligent Indian," as he wrote to his friend H. G. O. Blake. "Having returned," he added, "I flatter myself that the world appears in some respects a little larger, and not, as usual, smaller & shallower, for having extended my range."[1] Even without that clear, all-encompassing view from the summit, the trip was exhilarating, offering Thoreau a new perspective on his everyday horizons, the view of the world from his home in Concord. His challenge now, in the fall of 1857, would be to share that story in a literary travel narrative. That essay would become "The Allegash and East Branch."

In the final years of his life, Thoreau outlined a method of writing about lived experience that touched at the fundamental nature of human life on a changing planet. This book examines the development of the excursion as a literary method that veers away from more mainstream US travel writing toward an ecological worldview. Indeed, many of the narratives examined in this book were written in direct opposition to more mainstream literary travel narrative, and nearly all are relatively obscure today, generally read only by specialists in nineteenth-century literature. Thoreau is a notable exception, but beyond *Walden* and perhaps "Ktaadn," even his work has not drawn significant attention outside of specialist academic audiences. Nonetheless, these works are crucially connected to environmental writing today. For some nineteenth-century travelers, experiences beyond home presented a disturbance to more settled ways of understanding the planet. For authors who attended to these disruptions, such as Thoreau, Margaret Fuller, and the Ojibwe minister George Copway, nonfiction narrative offered a site of experimentation and adaptation, a place in which literature could respond to ecological ruptures ranging from deforestation and increased agricultural production to the breakdown of Indigenous ways of living in the world. In this sort of travel writing, the actual conditions of complex natural-cultural systems come to directly shape the ways in which a story is told. These patterns of disturbance and adaptation, disturbance ecologies, suggest a new form of resilience in which literary texts engage with and draw from the dynamic earthly conditions in which they are enmeshed. As such, these generally understudied and misunderstood works remain vital to the new literatures of the Anthropocene, the proposed and critically contested geological epoch in which humans have become a significant force on the planet's ecological and geological systems, with long-lasting, irreversible consequences.[2] The nineteenth century, much like the present moment, represents a notable threshold in the history of humans and the environment in North America and around the planet.

Many writers from the nineteenth century experimented with new ways of writing about the changing natural-cultural landscape, but Thoreau remains for many "the patron saint of American environmental writing," having emerged from relative obscurity in the nineteenth century to the status of a folk hero, albeit one who can still rankle students and readers today.[3] Nonetheless, this new way of writing took shape within a broader community of authors, and some of Thoreau's contemporaries developed their own distinctive yet related modes of writing about lived experience. Experimental writing, even when it breaks decisively from previous accounts of the landscape, requires models and precedents, and while he is often perceived as an isolated, even contrarian figure, Thoreau's work was part of a broader shift in literary form. To understand the experimental nature of his travel writing, this book thus examines several works by Thoreau in close detail. Along the way, I also pay significant attention to a handful of other contemporary authors of experimental travel writing, as well as one notable example of more mainstream travel writing, which appeared just as Thoreau was struggling to publish his own, more unsettling work.

Thoreau's writing has long been significant in the field of literature and environment, even as scholars have increasingly moved beyond nature writing to address a much wider variety of texts and forms. Lawrence Buell's landmark study, *The Environmental Imagination* (1995), treats Thoreau as a touchstone in American environmental thinking, remarkable both for his prescient representation of complex human-natural systems and the "fitful, irregular, experimental, although increasingly purposeful self-education" through which he came to better understand nonhuman nature.[4] Published the same year, Laura Dassow Walls's *Seeing New Worlds* (1995) established Thoreau's reputation as both a writer and a committed scientist, allowing him to develop an "innovative, experimental" style of thinking and writing "which combined transcendentalism with empiricism."[5] Since then, scholars have continued this work, drawing from new critical vocabularies and an increasing sense of urgency regarding climate breakdown, environmental justice, and ecological collapse. One important vein of scholarship addresses Thoreau's environmentalism and abolitionism and the intersections of environmental justice, embodiment, and race.[6] Another important and growing body of work explores phenology and environmental time, drawing from his Journal, the recently published Kalendar project, and other writings.[7] One particularly rich area of research has drawn from theories of New Materialism, exploring the role of matter in shaping texts from "Ktaadn" (1846) through the rest of his literary career.[8] Still, despite the range of this work, the various forms of Thoreau's writing, from his earliest travel writing to his late efforts to chart natural phenomena, remain understudied and not widely understood.

During his lifetime, Thoreau published or nearly finished four books: *A Week on the Concord and Merrimack Rivers* (1849), *Walden* (1854), *The Maine Woods* (1864), and *Cape Cod* (1865). Each of these works, and others by Thoreau, draws from the conventions of travel writing, but I am particularly interested in the last two, which are, for my purposes here, most unsettled and most reflective of an emergent literary practice, and, with *Walden*, most grounded in the close observations of the material world. As Buell notes, each of these four texts "centers on a relatively simple excursion or series of excursions," which he later expanded during the process of composition.[9] Moreover, despite comments in *Walden* about reading only "one or two shallow books of travel," Thoreau consistently engaged with both contemporary and historical examples of the genre (*Wa* 100). In his bibliography of Thoreau's reading in travel literature, John Aldrich Christie includes nearly two hundred titles.[10] Robert Sattelmeyer, in his foundational scholarship on Thoreau's reading and academic development, notes that this figure should be even higher, as it does not count New England travel writing and does not reflect works recorded in the previously unavailable Journal volumes of the 1840s.[11] This extensive reading directly influenced Thoreau's writing as well. Walls, in an essay on *Cape Cod*, "Walking," and planetarity, observes that his work "deliberately extends the genre of the scientific expedition to compel an *inner* exploration in tandem with the outer."[12] Interestingly, and despite this lifelong interest in travel writing, there have been no book-length studies of Thoreau and travel writing since Christie's *Thoreau as World Traveler* (1966). Furthermore, no research has directly connected this experimental travel writing to the emergent instability of the Anthropocene.

By tracing the direct links between lived experience and literary writing, and often between the experience of ecological rupture and the *failure* of established forms of writing to account for lived experience, this book outlines a process of disturbance and adaptation in which literary form responds directly to the instability of nineteenth-century environments and the resulting social shifts. Thoreau was not alone in grappling with the inability of mainstream travel writing to account for the realities of Indian Removal, industrialization, ecological collapse, and social inequality. Together, these (related) crises challenged a number of writers to reimagine their craft. These narratives deploy a structure built around a trip out and back, departing from and then returning home, where they can offer some entirely new sense of the world. These narratives would come to be grounded not in resort communities and new, fossil-fueled modes of travel but rather in disorientation, shock, and adaptation. Such stories remain vital because they suggest a method of writing for our own ecological moment, a means of revision and experimentation, and of tending to the fragmentation, disturbance, and adaptation that writers are still witnessing today.

This book focuses on literary craft, more specifically the construction of nonfiction narrative from lived experience and specific artistic practices. The methods of writing that gave rise to the excursion form resist theories of literature that insist on a rift between the text and the material world. In *Ecology Without Nature* (2007), for instance, Timothy Morton argues that no literary technique can overcome the distinctions between text and environment. Attempts at immersion, even when deploying first-person narration or extensive, skillful description, serve merely to further entrap the author and reader within the text. As Morton puts it, "The more the narrator evokes a surrounding world, the more the reader consumes a potentially interminable stream of opaque scribbles, figures, and tropes."[13] Alternatively, Buell uses Thoreau's *The Maine Woods* to outline what he calls a "nonfictionalist" theory of literature, which contends that "the emphasis on disjunction between text and world seems overblown."[14] Thoreau and other authors of nonfiction regularly weave the empirical and the literary into their work, suggesting a complex matrix of text(s), culture(s), and planet. Thus, Buell argues that literary texts have "a dual accountability to matter and to discursive mentation."[15] Although it runs against the grain of much contemporary literary theory, scholars have much to gain by addressing this potent contact zone of text, culture, and world. Indeed, to separate the world and the stories we tell about it seems to disregard the tangible, often destructive effects of human ideas on the landscape. In works of nonfiction in particular, readers must seriously consider the physical environment as a necessary precursor of writing and reading. In their departures and returns to our more familiar environments, excursions have the capacity to shift our sense of the world. As such, they offer a vital technique for environmental writing still today.

Thoreau's Early Excursions

The word "excursion" comes from the Latin *excurrĕre*, to run out. It evokes a deviation or a swerve, sharing a root with "excursive" and "excursus," which in the nineteenth century came to signify detailed discussions of some topic that digressed or was otherwise incidental to the main point, as in the appendix of a book.[16] Excursions are a departure and a return, and they might also take the form of textual diversions, narrative digressions, and ancillary points. Above all else, "excursion" refers to a short journey beyond one's recognizable, everyday environment, a veer, whether intentional or otherwise, off to the side of things. Throughout this book I use "excursion" to describe both the travels beyond one's home and back and the nonfiction narratives and essays that in various ways recount those trips. I also use "excursion form" to refer to the specific structure and narrative features that Thoreau developed in the final decade of his life. Excursion was a favorite word of Thoreau's, drawn from British Romanticism and used primarily to refer to his walks and longer trips. Of note, Thoreau's 1853 essay, "Yankee in Canada," was first titled "Excursion to Canada." He also used the word metaphorically, as in *Walden*, where he refers to giving "a true account of [life] in his next excursion" (*Wa* 91). Moreover, the first posthumous collection of Thoreau's essays was published in 1863 as *Excursions*, though he did not choose the title.[17] Even during his own lifetime, the term came to reflect the entanglement of nonfiction narrative and lived experience in the material world.

When Thoreau described the paradox of a new sense of the world, of things seeming "a little larger, and not, as usual, smaller & shallower" after his travels in Maine, he struck at the heart of a major change that was still taking place in the United States and its nonfiction literatures. Until the nineteenth century, American travel writing had been characterized by its wonder over American scenes and objects, its descriptive accounts of colonial-capitalist potential, and its analyses of the American continent, often rosy and propagandistic, but also frequently dark and defeated in the face of challenging social and environmental conditions. Wayne R. Franklin reads such texts in terms of three "essential forms of the travel book," each characterized by the author's relationship to the landscape, objects, and peoples being described. He calls these forms discovery narratives, exploratory narratives, and settlement narratives, with examples ranging from Thomas Jefferson's *Notes on the State of Virginia* (1785) to William Bartram's *Travels* (1791) and William Bradford's *Of Plymouth Plantation 1620–1647* (written 1630–51).[18] Franklin adds that these categories represent "ideal" forms, and that the varieties of travel writing often overlap and contribute to each other.[19] Often, individual texts, such as Bartram's *Travels*, might draw from key characteristics of, for example, both discovery narratives and exploratory narratives. These forms of travel writing might be considered not merely as aesthetic categories or narrative structures but also as orientations to the world, which would thus structure the ways in which American travelers both experienced and described landscapes across the continent. At the same time, many American authors read and drew from the writings associated with scientific expeditions, from Lewis and Clark's journals to popular narratives by Alexander von Humboldt and Charles Darwin.

By the 1820s and 1830s, forms of travel writing began to change, partly in response to the literary marketplace and partly due to the actual conditions experienced by travelers. As the nation found itself increasingly surveyed, mapped, and already described by previous generations of writers, readers at home and abroad still craved accounts of the American interior.[20] For many authors, this moment provided a key opportunity to break into a profitable literary marketplace, and the emergent tourism industry grew in tandem with a proliferation of guidebooks, travel essays, and other texts. Many canonical writers of the nineteenth century began their careers as successful travel writers before moving onto other genres. Still others, including Thoreau, would come to completely reimagine the shape, even the purpose of literary travel essays. For these authors, nonfiction literature offered a means of experimenting with narrative prose, of coming to terms with a rapidly changing United States. The proliferation of travel literature did seem to make the world a little smaller, a little stale, but, as Thoreau hints, the right kind of travel, and the right literary accounts of those journeys, could still make things fresh again, could even stretch one's horizons.

Many of Thoreau's earliest essays were travel narratives. His first work to reach a national audience, "A Walk to Wachusett," was published in *The Boston Miscellany of Literature and Fashion* in January 1843. The narrative recounts a three-day walk from Concord west to Mount Wachusett, about 30 miles away on the western horizon. His companion was Richard Fuller, whose older sister Margaret Fuller was then serving as editor of the Transcendentalist journal, *The Dial.* Two years before, Thoreau had submitted a poem addressed to the mountain, which she summarily rejected. In a letter dated October 18, 1841, she suggested a radical new approach: "I can have no advice or criticism for a person so sincere; but, if I give my impression of him, I will say, 'He says too constantly of Nature, she is mine.'" Nature, she went on, cannot be yours until "you have been more hers" (*Corr* 1:94). As Laura Dassow Walls observes, "A Walk to Wachusett" uses this approach, transforming "his pedestrian excursion into an epic mountaintop revelation."[21] As Thoreau and the younger Fuller approach the mountain, they observe a "truer and wilder pronunciation" of familiar place names, including "*Wor*-chusett" (*Exc* 33). After a night at an inn, they camp above treeline, sampling wild blueberries and reading Virgil and Wordsworth, observing what seems the whole of Massachusetts, spread out before them like a map. Reveling in all of this, the essay would offer something much more than a simple account of a walk, suggesting that these mountain summits offer intimations of "the general scheme of the universe" (*Exc* 42). In this, Thoreau hints at the revelations that will mark his later works. "A Winter Walk," published that same year in *The Dial*, shortly after Ralph Waldo Emerson had taken over as editor, also suggests Thoreau's later travel writing, offering what Walls describes as "a dreamy, evocative, intimate love letter to the tough New England winter" and the remarkable "metaphysical fire" beneath it.[22] Here too, travel writing offers a new sense of the world.

Thoreau's first book, *A Week on the Concord and Merrimack Rivers*, began as a travel narrative, built around a two-week excursion by water to the White Mountains in New Hampshire in 1839 and dedicated to his brother John, who'd died of tetanus in 1842. Writing in *Walden* years later, Thoreau muses that he has "woven a kind of basket of a delicate texture" that was "not worth any one's while to buy" (*Wa* 19). Most scholars see this as a comment on the structure and the eventual commercial failure of his first book. Linck C. Johnson, whose *Thoreau's Complex Weave* (1986) remains the most comprehensive study of *A*

Week, notes that Thoreau draws from two literary forms, the excursion and the elegy, plus a significant third strand, "the uses of the literary past," through which Thoreau comes to combine his account of the trip and its spiritual significance with a remembrance of his brother John and, finally, a detailed account of literature and writing from antiquity to the nineteenth century.[23] This final addition, the significant references to other literature, caused notable complaints among contemporary reviewers. Where readers expected a pleasant account of a river trip, with a climax atop New England's tallest peak, they found a long, meandering book, filled with digressions. Worse yet, those additions were a serious breach of good taste, a kind of scripture for the modern age. As Walls observes, Thoreau compresses his 1839 journey with John into a "creation story, a genesis of knowledge," a seven-day voyage toward Mount Washington, where the narrator ultimately offers only an Indigenous name, "Agiocochook," with no description (*Wk* 314).[24] Along the way, he puts the New Testament in direct dialogue with other religious texts from around the world. "Let but one of these sentences be rightly read from any pulpit in the land," he writes, "and there would not be left one stone of that meeting-house upon another" (*Wk* 73). *A Week* was, without a doubt, a radical experiment.

Emerson had urged Thoreau to publish *A Week*, and when the book failed to sell, Thoreau was financially responsible for six hundred unsold copies. Walls calls this the most "consequential event in Thoreau's life as a writer," a moment that would require nothing short of a self-reinvention.[25] In thus reimagining his work, Thoreau would recommit himself to direct engagement with the material world, beginning a process of making regular trips away from home and back again, which he records in his Journal and incorporates into everything that he wrote after 1850. This "turn" to the environment and to science has long been considered essential to *Walden*, a point I will return to in Chapter 3, but it has its roots in what Johnson calls "a series of personal crises and private revolutions" that shaped *A Week*, including Thoreau's "long and difficult struggle to assert himself as a man and writer" in relation to his mentor, Emerson.[26] To be sure, *A Week* is built around the narrative structure of an excursion, and it is doubtless experimental in terms of both its form and its content. Besides the extensive writing on world religions, Thoreau offers in his first book a watershed-based approach to the New England landscape and history, anticipating later modes of bioregional thinking.[27] Indeed, in connecting the mountains of New Hampshire and the low, grassy river valley of his home village in Concord, Thoreau not only reimagines the landscape but also confronts some of the changes caused by industrialization, including the construction of the Billerica Dam, which halted the migrations of anadromous fish and dramatically changed the hydrology of the Concord River valley.[28]

Ultimately, however, *A Week* would not contain what Buell calls "the representational density" of Thoreau's other later works. While his first book was revised primarily through the addition of "excursus meditations," the later three reflect the shift in Thoreau's Journal, which "became less a repository for thoughts, quotations, anecdotal vignettes, and drafts, and more a record of regular, meticulous daily extrospection."[29] In this process of reimagining his craft, Thoreau continued to experiment with literary structure, compressing two years in a single cycle of seasons in *Walden*, combining and drawing from several trips to the ocean in *Cape Cod*, and enacting three partially overlaying returns in three separate but crucially connected trips in *The Maine Woods*. In each of these three later works, Thoreau also drew not merely from memory, as in this 1839 excursion, but also from extensive field notes and the Journal. Thus, in these works, readers witness not only the development of an environmental consciousness and an image of an American literature grounded in the landscape but also a new *process* of writing about lived experience on a dynamic planet.

Disturbance, Adaptation, and the Excursion Form

In the late summer of 1857, H. G. O. Blake set out on his own solitary excursion, perhaps upset at not being invited to accompany Thoreau to Maine. Bradley P. Dean suggests that he likely hiked the now popular Crawford Trail in the White Mountains of New Hampshire, climbing over 4,000 feet to the summit of Mount Washington.[30] Later, he attempted to put his trip into words, and in November 1857, Thoreau wrote to him, "Let me suggest a theme for you: to state to yourself precisely and completely what that walk over the mountains amounted to for you,—returning to this essay again and again, until you are satisfied that all that was important in your experience is in it"[31]. Thoreau goes on:

> Give this good reason to yourself for having gone over the mountains, for mankind is ever going over a mountain. Don't suppose that you can tell it precisely the first dozen times you try, but at 'em again, especially when, after a sufficient pause, you suspect that you are touching at the heart or summit of the

> matter, reiterate your blows there, and account for the mountain to yourself. Not that the story need be long, but it will take a long while to make it short. It did not take very long to get over the mountain, you thought; but have you got over it indeed? If you have been to the top of Mount Washington, let me ask, what did you find there? That is the way they prove witnesses, you know. Going up there and being blown on is nothing. We never do much climbing while we are there, but we eat our luncheon, etc., very much as at home. It is after we get home that we really go over the mountain, if ever. What did the mountain say? What did the mountain do?[32]

In this call to write "precisely and completely," "returning to this essay again and again," Thoreau suggests both the difficulty of interpreting one's travels and, moreover, an author's accountability to what he called elsewhere "the *actual* world" (*MW* 71). *Cape Cod* and *The Maine Woods* are perhaps the best examples of Thoreau's mature excursion form, not because they are the least polished but rather because in their relatively unsettled state, at a remove from home, they better reflect Thoreau's very deliberate, relatively immediate efforts to make sense of the world. *Walden* too reflects this emergent form, though its descriptions and insights are carefully constructed from years of observation. Moreover, *Walden* is a peculiar kind of a travel book, unlike Thoreau's two later books in that it is as much an economic experiment and a memoir as a travelogue. Margaret Fuller's *Summer on the Lakes, in 1843* (1844) and George Copway's *Life, History, and Travels of Kah-ge-ga-gah-bowh (George Copway)* (1847) are also experiments with new means of accounting for the dynamic, unstable environments that the authors experienced, and can be productively compared to Thoreau's mature travel writing. All of these works feature experimental and intentionally unsettling techniques, such as digressions, fragmentation, shifts in genre, and breaks in readers' expectations. Many enact a deliberate departure from previous accounts of the landscape, setting themselves apart from the narratives of exploration and settlement that preceded them. Likewise, many look anxiously—and sometimes hopefully—toward an uncertain future. For all these authors, writing about travel offered a means of navigating a rapidly changing world.

By the nineteenth century, leaving home increasingly presented one with a strange, almost unrecognizable world, chaotic and difficult to navigate, offering unexpected challenges and also new charms. Because of an inherent sense of a departure and the novelty of their setting, excursions emerge from a cultivated attention to one's surroundings. As Dieter Schulz writes, excursions figure movement as a hermeneutical exercise, the narrator reading the world as they encounter it.[33] While all travel writing suggests a specific orientation toward geographic places, excursions often foreground an attempt to see differently from those who previously passed that way, to observe closely and offer readers some new perspective on a location that has already been described. In many cases, authors thus attuned themselves to a world that is in the midst of dramatic change. For author and reader alike, observations of such changes serve as a disturbance to more settled ways of looking at the world, a jolt or surprise that suggests a new outlook on the world writ large. To capture such experiences in literary prose, however, required a sense of traveling and writing as a new way of "apprehending from the thick of things," of capturing what Jeffrey Jerome Cohen and Lowell Duckert call "environment in motion."[34] In other words, the texts discussed in the following chapters offer a very particular sense of environment and environmental awareness, of ecologies shifting, coalescing, and taking on new shapes and functions in times of stress, uncertainty, and adaptation. Moreover, even as these works highlight the unrepeatable nature of lived experience—challenging the emergent tourist industry and the basis of much travel literature then and now—they also tended to highlight place and community rather than individual identity.

The term "environment," as Nicholas Royle puts it, centers on notions of the subject, of the observer, around which the environment turns. As Royle writes, "the 'vir' of 'environment' is from the French *virer*, 'to turn' or 'to veer' " and "at the center of this environing is, first of all, the I, the ego, the so-called individual and, by the same movement, the human."[35] In order for the environmental humanities to rethink "environment," Royle goes on, we must attend to the strangeness of the world around us. Cohen and Duckert, building on this, suggest recognizing and accepting "a world of inhuman forces, dynamic matter, and story-filled life that inevitably go off course," an attention to the way that these changes in direction enable "ontological, epistemological, and ethical positions to curl, converge, converse."[36] Less an organized field than a new intentionally divergent approach to ecological thinking grounded in New Materialisms and planetary dynamism, what they term veer ecology stresses the diverse human and nonhuman voices and energies that shape multispecies worlds. The texts examined in the chapters that follow anticipate this more modern awareness, deploying literary practices as a means of thinking through and adapting to emergent worlds. In the process, they raise important and vital questions. What can a narrative essay, an attempt at clarity or insight in literary prose, really *do*? If some surprising perspective or unexpected encounter can lead to a new insight into life back home, how can a writer create the right atmosphere for such wisdom to unfold? More fundamentally, how does a text demonstrate the openness and receptivity required for some new perspective to take shape? For twenty-first-century readers, excursions offer a sense of how we might navigate the instability of our present moment.

This book begins by examining the field of nineteenth-century travel writing more broadly to better understand the radical nature of the excursion form. Chapter 1 begins with Washington Irving's *Tour on the Prairies* (1835), which sets a precedent for a more subjective, innovative style of travel writing. Irving deploys a satiric, self-deprecating travel writing to "preserve" and thus fix a certain sense of the West. For Margaret Fuller, however, *Tour on the Prairies* came to serve primarily as a point of departure. In her *Summer on the Lakes*, Fuller draws from what I call "frontier optimism," the overwhelming sense of opportunity engendered by new possibilities, "vacant" land, and hope for the future in a period of dramatic political uncertainty and social and cultural change. When this sense of hope ultimately collapses, Fuller then attempts to gather up additional resources for her reader, critiquing them even as she does so. *Summer on the Lakes* is thus the first significant example of the excursion form, highlighting both the limits and the potential of this new way of writing.

Chapter 2 examines Thoreau's 1850 trip to Quebec, a failed experiment in literary tourism that led him to entirely reimagine his sense of travel writing. In "Yankee in Canada" readers see an emergent interest in instability as itself a form and a model for literary excursions, an idea that was radically at odds with mainstream travel writing of the day. George William Curtis, Thoreau's editor at *Putnam's Monthly*, would ultimately censor "Yankee in Canada," literally tracing out the limits of what an author might say about their travels. In the aftermath, Thoreau would continue to experiment with this material and with other narratives, and the excursion would develop in opposition to writing that drew from commercial tourism and mainstream, market-driven travel literature. While Curtis's own *Lotus-Eating: A Summer Book* (1852) frames travel and travel writing in terms of escapism and an aesthetics of leisure, Thoreau uses geographic movement as a means of exploring alternate histories and a failure of vision.

Chapter 3 considers Thoreau's "turn" to science and a more materially grounded style of writing drawn from lived experience. This chapter examines *Cape Cod* alongside *Walden*, focusing on what I call "environmental form" and the dual affects associated with disturbance and adaptation. In *Walden*, readers observe the startling revelation of earthly dynamism and, moreover, the narrator's new commitment to a literary process emerging from direct observation of the material world. *Cape Cod*, meanwhile, presents a darker affect and a more unstable sense of nature, economies, and US history. While *Walden* has been central to the field of ecocriticism and to Thoreau's reputation as an environmental writer, *Cape Cod* speaks directly to the Anthropocene, suggesting a method through which we can attempt to observe the larger systems shaping the world and our reactions to it.

The final two chapters consider two of the most radical examples of the excursion form. Chapter 4 examines Copway's *The Life, History and Travels of Kah-ge-ga-gah-Bowh*, which draws from the conventions of travelogues to place Indigenous narratives and the Ojibwe Nation within a rapidly changing world. *Life, History, and Travels* represents homesickness as a kind of constant companion, the narrative bending back toward Rice Lake, Ontario, again and again until finally, after a period as a missionary on the western shores of Lake Superior and a whirlwind tour of Boston and New York, Copway returns to his home ground to find everything changed. The concept of "home" in this text is ultimately not the same as for other authors, and it shifts over the course of the narrative. For Copway, the "good purpose" of a travel narrative becomes a matter of what George Vizenor calls "survivance," of holding onto and making use of an Indigenous identity in the face of a dramatically changing world.[37]

Chapter 5 takes up Thoreau's *The Maine Woods* (1864), a collection of three separate but crucially connected excursions published posthumously, each shaped by the presence of Penobscot voices. Each of these three essays charts a process of navigating and coming to understand something of the Maine North Woods, a region that fascinated Thoreau for its remove from Concord and Boston, but also for its relative proximity. Northern Maine was near at hand but a world away, disorienting in a way that profoundly shaped his larger sense of travel and the possibilities of travel writing. Ultimately, that sense of being lost would prove fundamental to Thoreau's writing, offering first a sense of earthly dynamism and, in the end, a dramatic reversal of those two foundational concepts for all travel and travel writing: "home" and "away."

In the coda, this book looks forward to the twenty-first century and Barry Lopez's *Horizon* (2019). Like earlier writers, Lopez deploys digressions, genre shifts, and disruptions to the formal and affective expectations of readers, aiming to shock or to disturb us into a new register of thinking and feeling. In doing so, he depicts literature as one part of a broader ecology of meaning, in which text and experience become part of an atmosphere in which wisdom can unfold. To read such a book, to consider how such a text is written, is to confront a question that emerges in the prologue, haunting Lopez's prose throughout—"What will happen to us?" In raising this question and circling back again and again, Lopez confronts the same sense of uncertainty that characterizes works by Fuller, Thoreau, and Copway. How can we move forward from a sense of crisis toward a better future? What can one *do* when the present seems dark and the future even darker? How can a writer work through the anxiety of tumultuous times, the uncertainty of a chaotic moment, to confront a changing world? The excursion emerged amid the specific conditions of the nineteenth century, including a literary marketplace that rewarded imaginative travel writing, but because these questions anticipate those faced by twenty-first-century authors, excursions can serve as an important starting point for studies of nonfiction literature of the present. Indeed, they remain essential to the new nonfiction literature of the Anthropocene.

1

"That West of Which I Speak": Imaginative Travel Writing

But we should do a great injustice to this book in representing it merely as a pleasant book of travels. It belongs to a class of which we can rarely find a specimen.

—James Freeman Clarke, Review of *Summer on the Lakes*

In *Summer on the Lakes, in 1843*, Margaret Fuller describes stepping off a steamship and onto an island shrouded in darkness. She is traveling alone from Milwaukee to Michilimackinac, where she expects to find a "large representation from the Chippewa and Ottowa [*sic*] tribes."[1] Before her steamer comes in, she casts these peoples as her fellow travelers, companions on their own "long holiday" (*Summer* 105). Then, just as she prepares to disembark, chaos ensues:

> As our boat came in, the captain had some rockets let off. This greatly excited the Indians, and their yells and wild cries resounded along the shore. Except for the momentary flash of the rockets, it was perfectly dark, and my sensations as I walked with a stranger to a strange hotel, through the midst of the shrieking savages, and heard the pants and snorts of the departing steamer, which carried away all my companions, were somewhat of a dismal sort though it was pleasant, too, in the way that everything strange is; everything that breaks in upon the routine that so easily encrusts us. (*Summer* 105)

In this moment, Fuller begins to see the West more clearly than at any previous point in the narrative. Ultimately, she will push against the limits of travel writing to suggest a different way of seeing and writing about this "new" place, but first she must help us, her readers, shake ourselves out of our own routine ways of seeing the world. What might one read, Fuller wonders, to have a truer sense of this region, these people? What would such a book look like? *Could* a work of literature break in upon that more settled way of looking at the world?

Since its publication in 1844, critics and reviewers alike have struggled to read *Summer on the Lakes* as a straightforward travel narrative, often dismissing it as unsuccessful or otherwise unworthy of serious attention.[2] Filled with veers, digressions, and poetic flights, it challenges us to keep up with a writer who is thinking on her feet. The book is, of course, a travelogue, but it is also a deliberate experiment in nonfiction narrative. To explore the changes brought about by this book, and ultimately its influence on Thoreau and others, this chapter begins by considering an imaginative turn in US nonfiction which took place nearly a decade before Fuller's time in the Great Lakes. While much earlier travel literature was judged in terms of accurate descriptions of routes and landscapes, nineteenth-century writers like Fuller increasingly found themselves describing places which were no longer "new," which had, in fact, been explored, described, and even settled by other authors. Published in 1835, Washington Irving's *A Tour on the Prairies* set a precedent for a more creative, literary style of travel writing. For Fuller, though, this romantic, cosmopolitan travel writing utterly failed to capture the confounding dynamism of the West, the dramatic veers that had characterized life in the region for two centuries, from the beginning of the French fur trade through the period of Indian Removal. She knew, in a word, that she needed to unsettle her readers, and she would go on to suggest a method for doing so.

Recognizing that established modes of writing and thinking were inadequate for describing Western environments, Fuller uses *Summer on the Lakes* to collect, transcribe, and critique as much as to simply narrate a journey. This experimentation would ultimately make the text something much more than a "pleasant book of travels," as her fellow traveler and reader James Freeman Clarke later wrote.[3] After crossing through a gateway to the West at Niagara, Fuller's reflections exhibit a tension between social critique and a utopian optimism—until she reaches Michilimackinac. There, finally unsettled by the enormity of Indian Removal, the narrative veers to ethnography and finally to a kind of bibliographic criticism. Fuller offers us her readings from Jane Johnston-Schoolcraft, George Catlin, Thomas McKenney, Chandler Gibbons, James Adair, Jonathan Carver, and others, arguing that each in turn is inadequate, leaving her unprepared to understand this place. To capture her journey, then, to account for the movement from settled life in New England to the Illinois prairies to Michilimackinac, she would need to speak directly to the archive itself, to the entire genre of nonfiction travel writing. In effect, the extensive swerves in this and other sections of the book, the digressions, extracts, and critical commentaries, come to operate as points of departure, shared resources and places from which to begin something new. Together, these moments represent a productive disturbance to nineteenth-century literary form, a moment in which nonfiction narrative must adapt to the dynamic conditions on the ground. *Summer on the Lakes* is ultimately an experiment not just in its style and form, as most scholars agree, but also in the efficacy of critique and sustained hope in uncertain times. As such, it is the first major example of what will become the excursion form, and for Thoreau, it proved a crucial model.[4]

The Imaginative Turn in American Travel Writing

When Fuller set out for the West, a broader transformation in American travel writing had, in fact, already begun. While earlier nonfiction travel writing had generally focused on the thrill of discovery, the wonders of exploring a "new" continent, and the challenges that arise in small, isolated communities, nineteenth-century travelers found themselves on familiar ground, and writers began to imagine new ways of telling stories about their travels. Indeed, Fuller was keenly aware of this, having read one of the authors who most successfully reimagined American travel writing. In May 1832, Washington Irving returned to the United States after seventeen years in Europe. He had written and published six books during that time, earning praise from critics on both sides of the Atlantic.[5] Still, he suffered acute anxiety over his reputation as an author, and as an *American* author, in particular. "But how was the wanderer to be received, after such an absence?" he muses in the introduction to *Tour on the Prairies*, his first book written and published after returning to the United States (*Tour* 9). The answer to this personal and professional crisis of identity, we come to learn, is in a "wide and varied" tour of the United States, of which Irving's time in the Indian Territory is only one part (*Tour* 10). By thus framing this narrative in terms of his own homesickness, Irving makes a shrewd professional move, capitalizing on the nation's obsession with the frontier while also drawing from the perceived associations linking remote territories, new beginnings, and American cultural identities. Touring the West, Irving reinvents himself while playing with the genre of travel writing. Along the way, he develops an engaging, imaginative style of travel writing, but he never quite embraces the most disturbing elements of travel.

When Irving set out for the Indian Territory in September 1832, he traveled with Henry Leavitt Ellsworth, one of three commissioners appointed by President Andrew Jackson to, in the words of Stanley T. Williams and Barbara D. Simison, "study the country, to mark the boundaries, to pacify the warring Indians and, in general, to establish order and justice" after passage of the 1830 Indian Removal Act.[6] Irving was further accompanied by the English writer and adventurer Charles Latrobe and the young Swiss Count Albert-Alexandre Pourtalès, as well as a mixed-race "braggart and a liar of the first order," Antoine, nicknamed Tonish, who served as "the squire, the groom, the cook, the tent man, in a word the factotum" and "the universal marplot" of the party (*Tour* 15). These companions made up what Commissioner Ellsworth called (exasperatedly) "the *Irving* party," who would travel together with Ellsworth's "newly raised corps of riflemen" (*Tour* 17).[7]

The trip became one of the most well documented frontier journeys in American literary history. In their journals, letters, and published accounts of the expedition, Latrobe, Pourtalès, Ellsworth, and especially Irving register an acute sense that the West is a place in flux, with dramatic change on the near horizon.[8] Lee Clark Mitchell writes that Irving kept his records first as a matter of "professional habit," his notes acknowledging "the transience of wilderness" and prompting him to "ensure an accurate account of what was quickly receding."[9] Interestingly, Mitchell argues that Pourtalès was the most sympathetic toward the plight of the Osages.[10] Irving, however, was uniquely focused on the changing nature of the region, "reverting again and again to the topic" of the passing West.[11] As he put it in a December 18, 1832, letter to his brother Peter, the region remained a "pristine wilderness," with Indigenous communities still living as "independent nations" and buffaloes not yet "driven beyond the reach of a civilized tourist."[12]

From the beginning, *Tour on the Prairies* deploys a rhetoric of preservation, which emerges amidst an acute tension between the mythic frontier with its "uninhabited" space and thrilling opportunities for adventure and, on the other hand, the experiences of traveling in the Indian Territory. Early in the book, Irving introduces the journey in terms of Ellsworth's role in Indian Removal, "an opportunity of ranging over those dangerous and interesting regions under the safeguard of a powerful escort" and surveying the exact areas "destined for the settlement of some of the migrating tribes connecting with his mission" (17–18). And yet it was the Osages, in particular, who most surprised Irving and enlivened his account of the West. These people prove, contrary to what he's read of Indians more generally, "by no means the stoics they are represented" (*Tour* 35). In his hands, they become, as Gerald Vizenor relates, "animate and humorous, not solemn, stoical, or heroic."[13] The West, and particularly the people that he meets there, are, from the beginning, surprising.

Just beyond Fort Gibson, the party encounters a small village of Osages camped on the banks of the Arkansas River. Irving describes "a number of old men" who "came forward and shook hands with us all severally" while women and children stood nearby, "chattering and laughing among themselves" (*Tour* 34). That night, when Irving and his party make camp nearby, several of them visit and mingle with the white travelers. Then, as they recline near the fire and accept gifts of food and coffee, Irving notes how these hunters "indulged in a vein of comic humour and dry satire, to which the Indians appear to me much more prone than is generally imagined":

> When they had made their supper and stretched themselves, side by side, before the fire and began a low nasal chaunt, drumming with their hands upon their breasts by way of accompaniment. Their chaunt seemed to consist of regular staves, every one terminating, not in a melodious cadence, but in the abrupt interjection huh! uttered almost like a hiccup. This chaunt, we were told by Beatte related to ourselves; our appearance, our treatment of them, and all that they knew of our plans. In one part they spoke of the young count, whose animated character and eagerness for Indian enterprize [*sic*] had struck their fancy, and they indulged in some waggery about him and the young Indian beauties that produced great merriment among our half breeds. (*Tour* 35)

Through Beatte, Ellsworth's mixed-race guide and by far the most experienced Western traveler in the company, Irving becomes aware of a joke that wasn't for him. Further, "Indian enterprize" seems to signal a broader joke, again at the expense of Pourtalès, about the way that white travelers expect to interact with Native peoples and their way of life on the frontier.

This moment demonstrates what was, for Irving and his white companions, an unexpected capacity for Indigenous joy. This joking at the expense of Pourtalès unsettles Irving's broader expectations of the West, too. *Tour on the Prairies* opens with a nod to previous descriptions of the frontier "from able pens," and yet literary writing has also left him curiously unprepared for this evening in camp (11). "In fact," Irving continues, "the Indians that I have had an opportunity of seeing in real life are quite different from those described in poetry" (35). Though sometimes taciturn in the company of whites, "when the Indians are among themselves" they are lively, observant, and humorous, "and entertain themselves excessively at the expense of the whites" (36). Vizenor concludes that Irving was moved by the humor and tricks of these Osages, and though none of these peoples are ever named within the text, *Tour on the Prairies* offers a crucial trace of Native character. As Vizenor puts it, "the action and reason he portrayed were closer to the literature of survivance" than nearly any other contemporary white travelers on the frontier.[14]

This surprising encounter with the Osages signals an early turning point in *Tour on the Prairies*, a pivot away from Irving's preconceptions of the West. If Irving and his companions are at first driven westward by their sense of a mythic frontier, by a chance to see the prairies before they are forever changed by US expansionism, settlement, and capitalist enterprise, this scene in camp suggests that the tension between expectation and lived experience, between the literature of the West and the scenes that Irving participates in, might give rise to some new means of writing and thinking about the remote spaces beyond the Mississippi. At the very least, it becomes clear in this moment that the travelers must in some cases adjust their expectations to the geography, the people, and the experiences they are presented with. Interestingly though, even as these long-held ideals about the West are challenged, the narrative continues to draw from and give rise to an enthralling sense of opportunity, perhaps a new kind of "enterprize," even as it foregrounds the dramatic changes unfolding on the landscape. Rather than merely ground truthing the various myths surrounding travel on the frontier, *Tour on the Prairies* will in fact recreate the exact conditions wherein Irving came to assess those myths in the first place, deploying the humorous, self-deprecating adventure tale to check his own status as a Western traveler and, furthermore, to ascertain the actual conditions of the frontier.

The stakes of this adjustment become most evident later in the book, when the reader becomes attuned to the same sense of imaginative possibility as Irving and his companions. Even while certain myths do not live up to actual frontier experiences, the environment continues to suggest thrilling encounters on the near horizon. The book's twenty-second chapter opens as the party makes an "indifferent encampment" and Tonish, a characteristically overeager and generally unsuccessful hunter, returns from a scouting trip ahead of the party. "According to his own account," Irving writes, "he had laid low two mighty bulls. As usual, we deducted one-half from his boastings; but now that he had something real to vaunt about, there was no restraining the valor of his tongue" (*Tour* 98). Then, after having "in some measure appeased his vanity," Tonish turns serious. During his daring exploit, he has also observed the fresh tracks of horses, which he concludes to "have been made by some roving band of Pawnees" (*Tour* 98).

Ellsworth, the official leader of the party, departs to reconnoiter the country ahead of them, making toward the bare summit of a nearby hill. The rest of the camp falls into a tense, uneasy routine, hobbling horses, gathering wood, and kindling fires, attempting to carry on as usual in the face of a coming catastrophe. Then, chaos breaks out:

> Suddenly there was an alarm of fire in the camp! The flame from one of the kindling fires had caught to the tall dry grass: a breeze was blowing; there was danger that the camp would soon be wrapped in a light blaze. "Look to the horses!" cried one; "drag away the baggage!" cried another. "Take care of the rifles and powder horns!" cried a third. All was hurry and scurry and uproar. The horses dashed wildly about—some of the men snatched away rifles and powder horns, others dragged off saddles and saddle bags: mean time no one thought of quelling the fire, nor indeed knew how to quell it. (*Tour* 98–9)

The quiet humor of the final line here, in which no one can think to extinguish the blaze, is almost lost in the passage's fast pace and confusion. It is finally Beatte, the only one in the group of Indigenous descent, who can ultimately control the blaze. Working "in the Indian mode," he and several others eventually manage to beat down the edges of the blaze with blankets. When the rangers join him in this act, the flames are at last "happily quelled" (*Tour* 99). The camp settles back into routine, the fires being kindled again, but this time in places where the dry grass is safely cleared away. All this time, no one has forgotten the coming Pawnees.

Tonish resumes cooking, promising a feast of fresh buffalo. Then alarm cries ring out again, this time from the far hill. Once again, chaos ensues:

> "The Captain has started buffaloes," cried one, "and wants horses for the chase." Immediately a number of rangers seized their rifles and scampered for the hill top [*sic*]. "The prairie is on fire beyond the hill," cried another. "I see the smoke—the captain means we shall drive the horses beyond the brook."
>
> By this time a ranger from the hill had reached the skirts of the camp. He was almost breathless and could only say that the Captain had seen Indians at a distance.
>
> "Pawnees! Pawnees!" was now the cry among our wild headed youngsters. "Drive the horses into camp!" cried one. "Saddle the horses!" cried another. "Form the line!" cried a third. There was now a scene of clamour and confusion that baffles all description. (*Tour* 99)

Amidst this moment of "clamor and confusion," Irving's narration bends toward a suspenseful immersion. Rather than listing and describing the events of the journey, the text recreates the same sense of uncertainty, the same profusion of possibility, that the party experiences, drawing from the techniques of a fiction writer to do so. What's important here is not the specific details, but rather the play of these particulars upon the minds of both the characters and the reader, who are suspended as though together within the story.

Surveying nineteenth-century travel writing, Lawrence Buell describes a turn away from description as a major transition in American travel writing, a shift attributable, on the one hand, to the influence of Romanticism and, on the other hand, to the increasing professionalization of literary marketplaces. *Tour on the Prairies*, like other works that would follow, thus displays "an interplay between the sequence of actual observations and the interests of a subjectively imposed mood or design."[15] The object, in other words, is not to merely portray the frontier and its potential for settlement, nor to offer a kind of guidebook for future travelers. Indeed, Irving is acutely aware that the West is so dynamic, these places so prone to dramatic change, as to preclude anything other than an act of preservation, of fixing that sense of adventure and possibility that had first attracted him to the region and which animates the party's departure from Fort Gibson. In his later Western books, *Astoria* and *The Adventures of Captain Bonneville*, he uses other genres and devices to fix the region's changing character in popular histories. His writing here, though, attempts to record and thus preserve the sense of adventure and delight that he is still able to find there.

While the cries of alarm ring out in the camp, Irving reflects on their position, noting the low hills that shut them in, occluding their vision and complicating the coming fight. He describes Ellsworth's soldiers, who up until now have been on what amounts to an extended hunting trip, enjoying what has been a high adventure rather than a military campaign. When Ellsworth returns to camp, they all gather round. He describes seeing a human form in the distance, "so perfectly motionless that he supposed it a bush" (*Tour* 101). Then, he goes on, another form appeared, and the two moved in a parallel fashion along the brow of the hill, stopping only when Ellsworth stopped. He *knows* that these were Pawnee scouts, observing the camp in preparation for an attack. Throughout the book, there is no moment so tense as this, when Ellsworth and his ranger look to "withstand siege" (*Tour* 102).

As the company reckons with their coming battle, another man on horseback approaches camp. A cry goes up: " 'It is one of the hunters'—'It is Clements!'—'He brings buffalo meat!'" (*Tour* 102). This wayward hunter has just had, by his own account, "a grand gallop" in pursuit of buffaloes. An incredulous crowd presses him:

> "Well, but the Pawnees—the Pawnees—where are the Pawnees?"
>
> "What Pawnees—?"
>
> "The Pawnees that attacked you—?"
>
> "No one attacked us."
>
> "But you have seen no Indians on your way?"
>
> "Oh yes, the two of us got to the top of a hill to look out for the camp, and saw a fellow on an opposite hill cutting queer antics, who seemed to be an Indian."
>
> "Pshaw! That was I!" cried the captain. (*Tour* 102–3)

And here, in Irving's words, "the bubble burst" (*Tour* 103). Realizing that there are no Pawnees, that they are in no imminent danger after all, the camp settles back into the usual routine. Another missing hunter, Old Ryan, has simply wandered beyond camp, and will return to the group after a night spent apart from the company, as he has done before. The thrill and the tension of an emergent conflict were imaginary all along, though no less stimulating as a result. Indeed, this fabrication and the moment when "the bubble burst" become a central motif. On the frontier, the reader realizes, the imagination runs wild.

Interestingly, this moment with the imagined Pawnees occurs shortly before the climax of the book, when the party finally reaches the Grand Prairie, where, to their "infinite delight," they emerge from a belt of timber and behold first a group of wild horses and then a pair of buffalo. This sudden appearance of characteristically Western game excites a new passion for the chase, but despite the building anticipation, the moment ultimately pales. First, Irving struggles to urge his horse near enough his quarry, and then his two pistols misfire. The hunt is unexpectedly challenging, too. "A gallop across the prairies in pursuit of game is by no means so smooth a career as those may imagine, who have only the idea of an open level plain," he writes (*Tour* 133). Rather, the prairies are difficult to traverse, and the buffalo he pursues descends into a narrow valley and escapes. Separated from his companions for a moment, Irving describes "something inexpressibly lonely in the solitude of a prairie" (*Tour* 134). Then he rejoins Pourtalès, who has also been unsuccessful. Together, they strategize.

Moments later, the two encounter an entire herd of buffalo. Afraid that the ensuing chase might leave them disoriented and far away from camp, they determine "to circumvent the herd, and by getting on the other side of them to hunt them in the direction where we knew our camp to be situated." Predictably, however, the herd takes alarm, and Irving and the Pourtalès pursue them in a full chase. Again, Irving struggles to urge his horse within range, and again his pistols misfire. Latrobe, suddenly on hand, lends him a double-barreled gun, and no longer needing to shoot from such close range, he singles out a bull and fires. The animal stumbles, and "the rest of the herd kept on their headlong career across the prairie" (*Tour* 136). Throughout *Tour on the Prairies*, Irving has anticipated this moment, but as the great animal falls, he experiences a rush of regret.

Approaching his "victim," he fetters his horse and reflects on what has just transpired:

> I am nothing of a sportsman: I had been prompted to this unwonted exploit by the magnitude of the game and the excitement of an adventurous chase. Now that the excitement was over I could not but look with commiseration upon the poor animal that lay struggling and bleeding at my feet. (*Tour* 136)

The animal continues to struggle, and Irving primes his pistols one more time, it now having become "an act of mercy to give him his quietus, and put him out of his misery." Thankfully, the weapon "for once proved true" (*Tour* 137). A moment later, standing over his prize, Irving measures the gulf between his expectations and the experience that has just transpired, between his emotions before and after the great animal fell. Ultimately, the thrill of the chase, the romance of a wild ride across the prairie, and the anticipation of it all fade in the rush of negative feelings that follow. This real, climactic moment, like the fight with the Pawnees, thus fails to live up to its imaginative associations, but rather than a bubble bursting in a moment of self-deprecating comedy, Irving is left "meditating and moralizing over the wreck I had so wantonly produced" (*Tour* 137).

Irving is not alone in voicing his regrets over this hunt. Pourtalès, for one, has suffered a great deal during the escapade. Immediately afterward, he is nowhere to be found. In fact, he's become separated from the group and will spend a night on the prairie apart from his companions, having ridden too hard in his pursuit and become lost. Later, he will recount camping in a grove atop a small rise, dreaming of his home in Switzerland. Only the next day will he be reunited with the party, after being found by the skilled trackers Beatte and Tonish. Commissioner Ellsworth, too, has suffered during the course of the hunt, riding a recently wild horse "with a considerable spice of devil in his composition" and killing a "huge buffalo bull" at close range, "almost in self defence" [*sic*] (*Tour* 139). Returning to camp "lauded on all sides" but also "grievously battered and way worn," he appears even more shaken by the experience than Irving:

> He turned a deaf ear to all compliments and congratulations; had but little stomach for the hunter's fare placed before him and soon retreated to stretch his limbs in the tent, declaring that nothing should tempt him again to mount that half devil Indian horse, and that he had enough of buffalo hunting for the rest of his life. (*Tour* 140)

With this taste of the Grand Prairie, seemingly all the party "battered and way worn," *Tour on the Prairies* turns back toward Fort Gibson. After they have hunted buffalo, pursued wild horses, and dreamed of encounters with Pawnee warriors, they seem to have exhausted the imaginative potential of the country all around them. Irving dedicates another chapter to observations of a prairie dog colony, a "village" or, in his eyes, a "republic," and then describes a council held in camp (*Tour* 145). The party had planned to travel further west, all the way to the Red River, but their horses are dangerously tired, having had little rest from their long journey and now having been ridden hard in pursuit of game. No one objects to a hasty retreat.

The final shape of this journey into Indian Territory is significant both for its literary style and also for its mediation of imagined possibilities and actual experience. In turning back, Irving indicates that accounts of the West must be more than simply a record of experiences, a description set down for future travelers. Rather, *Tour on the Prairies* recreates and then reflects on the thrill of novel experiences in distant lands. After the turning points with the Osages and the Pawnees, this final veer, both literal and figurative, suggests that the appeal of the frontier is largely in the imaginations of the traveler and the reader, the sense of possibilities and experiences out there on the near horizon. In Irving's hands, nonfiction thus entertains as much or even more than it informs, but it does not ultimately unseat or even destabilize those myths that first sent him into the West. Rather, the humorous adventure tale and, later, the Western history, including such works as his *Astoria* and *The Adventures of Captain Bonneville*, attempt to capture a fading or a vanishing West, to fix those myths for future generations.

Scholarship on Irving has consistently highlighted this sense of disappointment and surprise that emerges in *Tour on the Prairies*. Peter Antelyes, surveying

Irving's literary marketplace, describes an emerging connection between the spirit of enterprise, Western adventure narratives, and literary form. The origins of these "tales of adventurous enterprise," he writes, predate the colonial settlement of North America, but by the 1830s professional authors, Irving chief among them, would establish a crucial, though often uneasy connection between literary form and economic and territorial expansionism.[16] The "end result" of *Tour on the Prairies*, however, is a satire, as first Indian legends and then adventure narratives and then finally the idea of the West itself seem to fall flat in the face of actual experience: "The enterprising American had fooled himself into believing his own self-portrait as adventurer, and was failing to measure up."[17] Wayne R. Kime writes similarly, noting that by the end of this dashing adventure in the West, Irving came to see his account of the journey, and indeed himself, in a new light, not as a frontiersman, but rather as "a cultivated visitor from the East."[18] Other critics have also pointed to the tension that emerges between Irving's expectations and his actual experiences in the Indian Territory. Indeed, this sense that the West did not live up to his expectations is perhaps the most noted feature of the text. Mark K. Burns concludes that these "repeated absences, near-misses, delays, and disappointments" lead "the text's characters as well as the reader to reconsider, perhaps, the easy, transparent manner in which the American frontier was often thought to reveal itself and its significance."[19] James D. Lilley argues similarly, writing that "Irving will fail to capture events on the frontier not because his party just misses them but, paradoxically, because he is looking too hard for them."[20] *Tour on the Prairies* is remarkable, then, because of what it cannot quite replicate, because in its self-deprecating, often humorous critique of American adventurers, it suggests, above all else, that frontier travelers, setting out for the West with Cooper's *Leatherstocking Tales* or Philip Freneau's "Indian Burying Ground" and "Noble Savage" in mind, are critically unprepared to interpret and describe the scenes before them. What American adventurers need, Irving suggests, is an entirely new means of imagining the West. His mode of narration, of capturing the imaginative possibility of a specific moment in order to poke holes in his own status as an adventurer, would open the door for later writers of literary nonfiction. It would not, however, entirely unseat the tales of adventurous enterprise that it so ironically deflates.

More than a decade after Irving's trip to the Indian Territory, Margaret Fuller found herself in Chicago, passing time before she continued inland to the prairies. Describing that time in *Summer on the Lakes*, she reflects on reading "all the books I could find about the new region, which now began to become real to me" (*Summer* 19). She describes George Catlin's *Manners, Customs, and Condition of the North American Indians* (1841) as "far the best," though still not trustworthy. She admires Charles Murray's book, *Travels in North America in the Years 1834, 1835 & 1836* (1839), for its "accuracy and clear, broad tone," and for its originality (*Summer* 20). Of Henry Rowe Schoolcraft's *Algic Researches* (1839), she writes that "a worse use could hardly have been made of such fine material" (*Summer* 20). Anna Brownell Jameson's *Winter Studies and Summer Rambles* (1838) was better, though still too sentimental, and Thomas McKenney's *Sketches of a Tour to the Lakes* (1827) was "the dullest of books, yet faithful and quiet" with "some facts not to be met with elsewhere" (*Summer* 21). Irving's Western books, which Fuller read "some for the first, some for the second time," stand out amidst this bibliography. She goes on:

> Though the books are pleasing from their grace and luminous arrangement, yet, with the exception of the Tour to the Prairies [*sic*], they have a stereotype, second-hand air. They lack the breath, the glow, the charming minute traits of living presence. His scenery is fit only to be glanced at from dioramic distance; his Indians are academic figures only. He would have made the best of pictures, if he could have used his own eyes for studies and sketches; as it is, his success is wonderful, though inadequate. (*Summer* 21)

Fuller does not go on to describe the exact nature of this problem with Irving's histories. And while *Tour on the Prairies* stands out for the vivid experiences conveyed therein, it too has failed to leave her prepared to see the West, as she will find.

In Fuller's overview and critique of Western writing we find both the promise and also the shortcomings of Irving's more imaginative style of travel writing. In mediating myth and experience, possibility and disappointment, *Tour on the Prairies* ultimately set out on a project that it cannot complete. *Tour on the Prairies* creates the conditions wherein certain experiences of the West are preserved through literary writing, and while the book casts its narrator as a somewhat green traveler, it ultimately suggests that such possibilities *might* still be out there, that a certain kind of masculine adventure remains characteristic of Western experience. While Irving set a precedent for a new style of travel writing, for nonfiction that would reframe the West in terms of imaginative possibility and the inherent conflict between experience and literary narration, his satiric, self-deprecating form never fully confronts the most unsettling aspects of social and ecological change in the region. For Fuller, the dissolution of these myths would prove crucial, and capturing her experience of the West would involve more than an immersive, imaginative style of writing. Writing about Illinois and Wisconsin and Michilimackinac, she came to see, would require dramatic veers and polyvocal accounts of places in flux. Reading Irving, and particularly *Tour on the Prairies*, Fuller came to see that in order to reinvent travel writing as a literary form, she must first cast off from the genre itself.

Summer on the Lakes and the Excursion Form

Summer on the Lakes begins at Niagara on June 10, 1843, where Fuller encounters the famous falls as a portal to what seems another world, a gateway from western New York through the Great Lakes and ultimately to the prairies of Illinois. She knew that for many Americans, the trip from Buffalo to Chicago, from Lake Erie through Lakes Huron and Michigan, represented the most direct track toward a better future. Fuller reflects on Niagara and her route: "There can be no two places in the world more completely thoroughfares than this place and Buffalo. They are the two correspondent valves that open and shut all the time, as the life-blood rushes from east to west, and back again from west to east" (*Summer* 19). But does the West, she muses, live up to everything that she's read? Is it really a region defined by freedom and opportunity for all? Standing at the edge of Niagara Falls, she comes to see that she is not only unprepared to see the West, but that she has been moving toward an illusion. While *Summer on the Lakes* plays with a sense of frontier optimism, the book is grounded in critique, exploring the ways in which nonfiction literature might contribute to broader projects of reform. Even at Niagara, at the very beginning of the journey, Fuller struggles with fundamental questions about the purpose of travel writing in a changing world. Ultimately, these questions will emerge as the central theme of the narrative.

Fuller begins the book by emphasizing the fundamentally unbridgeable gulf between writing and living, between the page before us and the actual sensation of standing near Niagara Falls. Ultimately, this will lead her to address the very reasons for traveling to the West in the first place:

> Since you are to share with me such foot-notes [*sic*] as may be made on the pages of my life during this summer's wanderings, I should not be quite silent as to this magnificent prologue to the, as yet, unknown drama. Yet I, like others, have little to say where the spectacle is, for once, great enough to fill the whole of life, and

> supersede thought, giving us only its presence. "It is good to be here," is the best as well as the simplest expression that occurs to the mind. (*Summer* 3)

By invoking "foot-notes" on "the pages of my life," Fuller establishes the intimate, eclectic form of the book, in this case a memoir and a narrative, but also the notes, fragments, and connections that might have existed below the margins or off to the sides of a more straightforward narrative or a guidebook. This opening sentence also establishes a certain sense of privacy even amidst these notes, "an unknown drama" that will still never quite come into relief. Most immediately, though, this passage suggests that even a critic and a writer such as Fuller might be stilled by some overwhelming natural object. In the presence of some vast natural feature of the landscape, language and even thought itself fade, becoming inadequate to the project of description, much less narration. Describing such sensations, we might infer, is not merely impossible, but somehow counterproductive. *Summer on the Lakes* promises an account of a journey from Niagara to the prairies of Illinois and back again, but even in its opening lines the text gestures toward the impossibility of capturing any enduring truth about the West, or even experience itself.

Ultimately, Fuller finds that "magnificent prologue," Niagara Falls, somewhat disappointing. She has, in a way, already been here in "drawings, the panorama, &c." and, she goes on, these have "given me a clear notion of the position and proportions of all objects here." As she puts it, "everything looked as I thought it would" (*Summer* 4). Even by 1843, Niagara Falls had become a major destination for tourists. Fuller thus finds her experiences, even her reactions, scripted. In the words of Robert E. Abrams, "such an experience of place is already belated even on arrival," the scene "so standardized in image and print" that Americans arrive knowing exactly what to expect.[21] Worse yet, at Table Rock, near the base of the falls, Fuller observes a fellow traveler spitting into the rushing water. Moments like this, she writes, will hopefully not "be seen on the historic page to be truly the age or truly the America" (*Summer* 5). Are American travelers, she wonders, much less settlers, fit to experience the West, to properly observe this landscape? If not, what sort of literature or art might prepare them to see, to describe this new world? Fuller will return to this question throughout *Summer on the Lakes*.

When Fuller departed for the West, she carried with her a certain sense of impasse. On one hand, years of intellectual labor had left her exhausted, and travel offered a much-needed respite. She'd edited the Transcendentalist journal, *The Dial*, from its beginnings in 1840 until the spring of 1842, but was never paid the salary that Emerson had promised her at the outset. Moreover, *The Dial* was faltering, unable to attract more than a handful of subscribers and regularly mocked by reviewers outside of Emerson's literary circle. She seemed to have hit a dead end in her other work, as well. Since 1839 she'd hosted what she called Conversations, gathering small groups of women in parlors to discuss the classics and the "great" questions. When she attempted to expand her audience to include Emerson and other men, however, the dynamic changed and the series came to an end.[22] In the words of Caroline Healey, a teenager who was in the room when Emerson was first in attendance, Fuller "never enjoyed the mixed class, and considered it a failure so far as her own power was concerned."[23] Fuller had also just finished a long, arduous project on the condition of women in her contemporary United States, "The Great Lawsuit. Man versus Men. Woman versus Women."[24] Furthermore, she had just completed an English translation of Bettina von Arnim's *Günderode*.[25] Put simply, she was suffering from exhaustion, and though she'd dreamed of a yearlong tour of Britain and Europe, a four-month excursion in the West, from Niagara Falls to Chicago and Sault Sainte Marie and back again, would have to do.

At the same time, another, less personal predicament was looming. Against the backdrop of an increasingly energetic abolition movement and a rapidly industrializing north, Fuller and her contemporaries expressed a growing concern over the future of the United States. Reading Thomas Carlyle's 1839 assessment of England under capitalism, *Chartism*, Fuller began to muse on the "Condition-of-England Question," wondering what an industrialized, capitalist United States might come to look like in the years ahead. When she set out for Niagara in the early summer, she packed Carlyle's *Past and Present*, which prescribed authoritarian rule as the only means of countering an increasingly large and increasingly unruly working class, ultimately arguing that only some combination of class hierarchy and religious leadership would be able to maintain order and preserve economic and social functions. This rankled Fuller.

On June 1, Fuller wrote from Niagara to Emerson, who had helped publish Carlyle's book in the United States only a month after it was released in England.[26] Even as she admired the analysis of social class, she complained that Carlyle outlined "no valuable doctrine" which might address the wicked problems of the (modern, industrial) nineteenth century:

> His proposed measures say nothing. Educate the people. That cannot be done by books, or voluntary effort, under these paralyzing circumstances. Emigration!

> According to his own estimate of the increases in population, relief that way can have very slight effect. He ends as he began; as he did in Chartism. Everything is very bad. You are fools and hypocrites, or you would make it better.[27]

With this book in the back of her mind, Fuller's account of the journey West unfolds as a search for answers about life in industrial modernity. *Summer on the Lakes*, as Lance Newman puts it, "is Fuller's response to the Condition-of-America question," her meditations on the future occasionally eclipsing the book's account of an actual journey.[28] If she first traveled west to seek respite from the personal and professional tensions that she faced in New England, it quickly became apparent that on the frontier, the question of America's direction loomed large. For an exhausted editor, the West would indeed offer a kind of hard reset, a much-needed change of scenery, but it would also throw questions of environmental determinism, landscape aesthetics, social progress, and Indian Removal into new relief.

From this sense of impasse, *Summer on the Lakes* moves forward into a "new" world, veering between narration, reflection, and critique. Even if it can't offer the kind of proposals that Fuller had hoped for in reading Carlyle, *Summer on the Lakes* attempts to model a means of working through what it does know: the conditions on the ground, the voices present in the room, and the way that it is all contingent and malleable, produced in real time as conditions in the West continue to change. Throughout, but especially in her descriptions of northern Illinois, Fuller draws from frontier optimism, a kind of culturally pervasive hopefulness over the prospects of the West which is grounded in dreams of a better world. Unlike many frontier writers and thinkers, though, her engagement with the West as an idea and as lived experience is both critical and utopian, progressive even as it assesses the prospects for lasting social change in the region and across the country. It draws from a ubiquitous cultural affect—a sensation, really—even as it gauges the possibility of real improvement. Fuller also confronts the limits of this optimism, the point at which the West or any kind of cultural attachment becomes a "utopian fantasy" rather than a real opportunity. Where is the line between an idealist's reverie and a pragmatist's critique, and how might we come to mark out that space between the two extremes? Moreover, what can Fuller's encounters with the limits of her own hopefulness offer contemporary readers? By focusing on critique and experimentation, *Summer on the Lakes* offers both a methodology and some attempts at answers.

As Fuller travels inland from Chicago into the Rock River Country and observes the tallgrass prairie, her descriptions veer from an initial aversion to such big horizons to a utopian vision of American possibility—on the frontier, she muses, there is time and space for both a hearty agrarian lifestyle *and* the intense self-cultivation that she so values. This swerve has been the subject of debate among critics. On one hand, Annette Kolodny reads the Rock River chapter as a sentimental reverie, a lapse in the book's otherwise sharp critical judgment. As she puts it, this hopeful interlude merely dreams of "the unmediated pleasure" of Fuller's occasional childhood escapes to her mother's garden, where she could be free of her father's rigorous standards of self-culture.[29] Conversely, Lance Newman argues that *Summer on the Lakes* responds directly to Carlyle, and the Rock River chapter thus engages the "Condition-of-America question." As he puts it, this chapter is not "a lapse into nostalgic reverie" at all, but rather a social critique that "uses the idiom of the picturesque to envision an organic utopia, an alternative to the steady expansion of capitalism."[30] That is, in her descriptions of the prairie, Fuller imagines an alternative to Carlyle's sooty industrialism and appeal to authoritarianism.

This debate over Fuller's interpretation of the Rock River Country tends to occlude the real stakes of her observations, however. Kolodny's emphasis on accuracy and clear-headedness seems to disregard Fuller's persistent hopefulness, and Newman's focus on the text's underlying critique of Carlyle does not quite capture the text's dramatic searching for answers. Even as *Summer on the Lakes* offers its rosiest descriptions of frontier life, Fuller realizes that it is not quite adequate, that she can't quite linger in that state of blind confidence over Western expansion, just as she knew that she could not quite *settle* at a utopian community like Brook Farm.[31] In other words, Fuller responds to Carlyle even while knowing that she cannot quite address the condition of America—at least not yet, with only a collection of observations of this nascent, Western society. The book can only, to borrow Lauren Berlant's words, gather up its resources, preparing in some way to deal with the crisis of living in what was rapidly becoming an industrial-capitalist nation.[32] It suggests that when we face an impasse, accuracy and precision and critical thinking can only go so far. The Rock River chapter is thus not simply an exercise in imagining utopia, but an assessment of that leap of faith, a measure of what it might take to live through the transition. It does not naively envision some utopia arising from the Midwest soil but takes stock of frontier life to imagine what it might take to build a livable world—and its future thinking draws explicitly from her observations of frontier women and Indigenous people.

West of Chicago in the Rock River Country, Fuller begins to imagine the West in terms of some distant, but not unforeseeable future. Here, she remarks that the landscape seems like a garden, a place "which has been inhabited by a nation skilled like the English in all the ornamental arts of life, especially in landscape gardening," except unlike in England, there seem to be no borders, no enclosures. The landscape appears, rather, a kind of Jeffersonian ideal. Fuller describes it as a model for Midwestern society, a "pleasant society" forming among the diverse group of new settlers, who display "cultivated minds and refined manners" (*Summer* 38). In treading this ground, Fuller might have simply modeled her ideal republic and moved on, content to note how the West *could* move forward with a sense of itself as a kind of agrarian idyll. In doing so, however, she confronts, in Newman's words, "the oppressive reality of America's developing social order" at a broad scale, including a specific kind of sexism.[33] The future of this "pleasant society" is critically limited by the gender roles assigned to women, "to whom a city education has imparted neither the strength nor the skill now demanded." Fuller calls this the "great drawback upon the lives of these settlers," the single largest hurdle facing her ideal society (*Summer* 38). Whereas men are afforded all of the opportunities and charms of a "new life" on the frontier, the women are faced with a greater share of the labor and relatively fewer rewards.

Despite this, *Summer on the Lakes* continues to draw from a general sense of frontier optimism. Even after putting her finger on the problem of women's frontier lives, she *imagines* a frontier woman, someone who arises from the Midwest soil to address the changing nature of American life. The girls that she sees along the Rock River and its tributary, the Kishwaukee, might someday escape their present "unfitness." Fuller imagines such a frontier woman: "An elegance she would diffuse around her, if her mind were opened to elegance; it might be of a kind new, original, enchanting, as different from that of the city belle as that of the prairie torchflower from the shopworn article that touches the cheek of that lady in her bonnet" (*Summer* 39–40). Here, in short, are the stakes, the sense of opportunity, but also the challenge—if the Illinois prairies can give rise to a generation of women who are fitted to the place, who are truly a product of the Tallgrass Prairie and the Rock River watershed, the Midwest might yet be a certain kind of utopia, a place where, in Fuller's words, "all have in common the interests of a new country and a new life" (*Summer* 38).

This moment on the tallgrass prairie, when Fuller identifies gender as the most significant obstacle to her idealized society, marks the point at which her various intellectual energies converge in a single, recognizable literary voice. Kolodny writes that *Summer on the Lakes* represents a turning point in women's accounts of the frontier, Fuller's work building on "The Great Lawsuit" and *Woman in the Nineteenth Century* but also "informed by the new insights (and the frustrated fantasies) awakened on the prairies."[34] Similarly, William Stowe argues that the polyvocal approach she pioneers in *Summer on the Lakes* later comes to fruition in her letters from Italy, and that the book therefore played a central role in her development of a distinctive voice and critical approach.[35] Margaret Allen refers to this moment in more explicitly political language, calling it the beginning of Fuller's transformation "from political innocent to dedicated social activist."[36] Newman simply calls *Summer on the Lakes* and the journey west the beginning of "a process of rapid and decisive radicalization."[37] This sense of frontier optimism—and also its limits, which she will confront still more directly on Michilimackinac—was an essential building block for Fuller's lifelong intellectual project of imagining and working toward a more just, intellectually and culturally richer future for everyone, and for women in particular.

When Fuller departs from Illinois, she travels north to Milwaukee, then a small, newly established city on the oak savannahs of southeastern Wisconsin. On a brief tour inland, Fuller and her companions come upon a band of Potawatomi camped on the banks of Silver Lake, about 25 miles to the west. When a sudden thunderstorm comes up, the travelers are forced to seek refuge. Fuller, reflecting on her discomfort, describes crowding into their lodges:

> These [lodges] were very small, being for temporary use, and we crowded the occupants much, among whom were several sick, on the damp ground, or with only a ragged mat between them and it. But they showed all the gentle courtesy which marks them toward the stranger, who stands in any need; though it was obvious that the visit, which inconvenienced them, could only have been caused by the most impertinent curiosity, they made us as comfortable as their extreme poverty permitted. (*Summer* 74)

Recently displaced from the area by an influx of immigrants, some of whom harken back to the idealized settlers of northern Illinois, these wandering Potawatomi "had returned on a visit, either in homesickness, or need of relief" (*Summer* 75). The moment is the beginning of an awakening for Fuller, and it will engender yet another veer in the text.

Earlier, in the Rock River Country, Fuller connected her idealized settlers to the recent presence of Indigenous peoples, drawing an aesthetic link between these agrarian homesteaders and the recently displaced Potawatomi, Miami,

Sauk, Fox, Ho-Chunk, and others. Some settlers' houses, she writes, "showed plainly that they had no thought beyond satisfying the grossest material wants" (*Summer* 29). The idealized settlers, however, live in a certain harmony with the landscape, drawing from Indigenous cultures as they learn to blend into the unique environment of the Tallgrass Prairie:

> Seeing the traces of the Indians, who chose the most beautiful sites for their dwellings, and whose habits do not break in on that aspect of nature under which they were born, we feel as if they were the rightful lords of a beauty they forebore to deform. But most of these settlers do not see it at all; it breathes, it speaks in vain to those who are rushing into its sphere. Their progress is Gothic, not Roman, and their mode of cultivation will, in the course of twenty, perhaps ten, years, obliterate the natural expression of the country. (*Summer* 29)[38]

Fuller's sense of the Rock River Country did not include Indigenous peoples, and neither in Illinois nor in Wisconsin would she go so far as to critique the broader project of Indian Removal, but her utopian reverie will also not spare the settlers who most grossly exploit Indigenous lands. Indeed, Fuller would assign proper ownership of Rock River, at least in an aesthetic sense, to these former inhabitants, the "rightful lords" of the land she observes. *Summer on the Lakes* cannot imagine a continued presence of Indigenous peoples in northern Illinois or southern Wisconsin, but just the same, the text cannot figure a version of the landscape where those peoples have entirely vanished either.

Later, but still before her encounter with the Potawatomi family, Fuller had encountered a young Indigenous woman who shares the medicinal virtues of some of the flowers then in bloom. Fuller, impressed, reflected that the Indians, "those students of the soil," possessed an intimate knowledge of plants that is lost on settlers and travelers, who "could only look to admire its hues or shape" (*Summer* 41). In her admiration Fuller expresses in these lines the seeds of an Indigenous environmentalism and gestures toward the importance of Indigenous knowledge. By deferring to the far superior experience of this young girl along the Kishwaukee River, she strengthens her previous link not only between Indigeneity and landscape aesthetics, but also to ecological well-being. Considering the treatment of the land by settlers, she muses that this natural "garden" will soon fall victim to introduced species, prairie flowers giving way to "the coarser plantain, the 'white man's footstep'" (*Summer* 41). By the time she meets the Potawatomi camped along Silver Lake, she is primed to see the plight of Native people as a result of settlers' attitudes and values, but she still holds back from critiquing the popular racist metanarrative of "fading" or "vanishing" cultures. Only when she arrives on Michilimackinac does Indian Removal take on an arresting sense of immediacy, becoming the single most dire problem of the West.

Setting off alone into the darkness when she first arrives on the island, Fuller's own routines, indeed her plans, were quickly disrupted. Although expecting a room to herself at the hotel, she finds none and is obliged to stay in the common parlor and dining room. In the days that follow, she lets these circumstances guide her impressions. She was, without a doubt, disturbed, shaken out of her routine ways of thinking and being in the world. If the prairies and the oak savannahs engendered a certain utopian thinking, the northern Great Lakes will leave her entirely unsettled. The island, she writes, is beautiful, "a scene of ideal loveliness." She is particularly impressed by the "arched rock," which surprised her, "much as [she] had heard of it, from the perfection of its arch" (*Summer* 106). Above all, though, it is her fellow travelers, the Ojibwe and Odawa encamped on the beach, who most impress her, most animate the account of her nine days on the island. Like Irving on the prairies, she is touched by the humanity of these fellow travelers. Communicating via signs and reflecting on the plight of Native women in comparison to the wives of white settlers, Fuller is genuinely moved, struck by the animated speech, the variety of gestures, the delicacy and decorum she observes on the beach. And yet, this is no place for optimism, either, for creating a shared dream of a new society. Rather, it becomes a sobering veer toward Fuller's own culture, and, indeed, the failure of the settler imagination.

Rising before the other hotel guests each morning, she joins her "Indian neighbors" near the water's edge, communicating with them through an improvised sign language. Her account of these moments evinces an unusual sympathy, but it quickly moves from ethnography to something else, a kind of future-oriented social criticism. Having encountered the young girl along the Kishwaukee and sheltered with the band of Potawatomi along the banks of Silver Lake, Fuller seems to finally reckon with the enormity of Indian Removal, with the social and ecological disturbances resulting from US settler colonialism, her idealized Rock River Country not excluded. Here, in the longest chapter of *Summer on the Lakes*, Fuller turns from narrative to bibliography, offering us her readings from Jane Johnston-Schoolcraft, George Catlin, Thomas McKenney, Chandler Gibbons, James Adair, Jonathan Carver, and others. She inserts among her own reflections a speech by the Massachusetts Governor Edward Everett to a Sauk and Fox congregation in Boston, as well as an unattributed Native tale that had, as she informs us, never before appeared in print, "Muckwa, or the Bear." Drawing from all these readings, Fuller presses on white travelers' lack of sympathy, clearing space for a new treatment of Indigenous peoples in literary texts. She holds back, though, from imagining an Ojibwe or Odawa future. Indeed, she admits that she cannot: "I have no hope of liberalizing the missionary," Fuller writes, "of humanizing the sharks of trade, of infusing the conscientious drop into the flinty bosom of policy, of saving the Indian from immediate degradation, and speedy death" (*Summer* 121). Fuller uses every literary resource available to her, because this is, she realizes, a problem of imagination and of narrative.

In considering the fate of Native women, Fuller offers a blistering critique of an ideology that scholars now call savagism, which centered around the racist belief that Indigenous peoples were doomed to vanish in the face of encroaching "civilization."[39] Although she suggests that the loss of Native cultures is imminent and perhaps irreversible, as in her insistence on "a national institute, containing all the remains of the Indians," Fuller's stance is quite clear (*Summer* 143).[40] By this late summer morning on the beach, she has had nothing short of an awakening. As Susan Belasco Smith puts it, "Fuller was appalled by the systematic removal of the Indians from their ancient territories," and thus "admonishes her readers with periodic lectures on the subject."[41] In fact, her critique is harshest toward white women, perhaps because she senses how hard she has had to work to rid herself of her own prejudices:

> I have spoken of the hatred felt by the white man for the Indian: with the white women it seems to amount to disgust, to loathing. How could I endure the dirt, the peculiar smell of the Indians, and their dwellings, was a great marvel in the eyes of my lady acquaintance; indeed, I wonder why they did not quite give me up, as they certainly looked on me in great distaste for it. (*Summer* 113)

These "lectures" are more than a mere critique of settlers' shortsightedness and prejudices. On Michilimackinac, Fuller realizes that issues of gender and race are fundamentally intertwined with the future of specific geographic places, and she thus confronts both the difficulty and the critical importance of what we today call intersectionality and environmental justice, in this case a place-based, reform-focused writing which combines feminism and social justice with a focus on environment. Still, Fuller cannot ultimately solve the crisis of Indian Removal, nor the related crisis of gender oppression which she came to see on the Illinois prairies. The book's veers thus become, as Cheryl J. Fish writes, "indicative of the difficulty of formulating a discourse of inclusiveness whereby women and other oppressed groups could act together in solidarity."[42] In attempting to work through these related crises, Fuller can only offer a creative-intellectual methodology. Just the same, *Summer on the Lakes* finally aims to wake its readers up to the problems of Indian Removal and to the dangers of other Western fantasies, and all of these veers are in answer to that question of form, of how to break in on a more settled way of viewing the world.

Charles Capper writes that *Summer on the Lakes* has "provoked a good deal of critical puzzlement," partly over its "comments about women and Indians," and partly, too, because of a certain "impatience with its interlaced musings, dialogues, tales, poems, book critiques, and other seeming digressions."[43] But these "interlaced" veers prove crucial to the book's success in imagining a better world. Ultimately, the book aims to capture Fuller's means of thinking through these wicked problems, suggesting in the process a methodology for travel writing in a dynamic, dangerous world. As Fish argues, the book is capable of "transcending the travelogue through its blend of autobiography, history, critical reading, and a gender and race-based analysis of the promise and tragedy of westward expansion."[44] Nicole Tonkovich suggests that it is "a synecdoche of the library itself, an artifact marking the difficulties facing women writers of Fuller's education and social position at the beginning of literary professionalization in the nineteenth century."[45] In other words, the book measures and assesses the very possibility of capturing this sort of awakening in literary prose. This calculated attempt at capturing the depth of one's personal journey, the strategic refusal to follow generic forms, makes *Summer on the Lakes* a crucial model for later excursions, for accounts of disturbance in which the traveler comes home shaken up, unsettled, and unable to look at home in quite the same way.

Like *Tour on the Prairies*, *Summer on the Lakes* sets out with a certain generic expectation, promising a Romantic tour of the Great Lakes with an extensive commentary on landscape aesthetics. On the prairies, it swerves toward a certain kind of settlement narrative, a cautiously optimistic sense of what might become of the region we now call the Midwest. Ultimately, this final veer then collapses on the beaches on Michilimackinac, and we must journey toward home again—but this journey has not been for nothing. In the seventh and final chapter, Fuller describes shooting the rapids at Sault Sainte Marie, but after those days on the beach, it seems little more than a diversion with Native guides. Boarding a steamship bound for Buffalo, the book closes abruptly a scant few pages later. Unable to fully endorse any single account of the frontier, she thus leaves us with the suggestion that we need some other writer, an Indigenous author, to account for this sense of the West, for the sensations that she has come to know. Faced with the task of providing a new means of writing about travel, of accounting for the *actual* West, Fuller ultimately offers a reading list and a wave. Then she departs again, leaving us as she heads for home.

Fuller finished writing *Summer on the Lakes* on her birthday, May 23, 1844, less than a year after she returned from the West. The moment came to mark a decisive turn in her career as a writer, critic, and journalist. That autumn she moved to New York, where she joined the staff of Horace Greeley's *New York Tribune*, becoming the first woman in the United States to work as a professional literary critic, editor, and, by 1846, foreign correspondent. Over the next four years she produced some 250 pieces for the *Tribune*, most of them with the byline "*".[46] During that time, she would dedicate herself not only to book reviews, but to front-page columns on serious social issues, often putting herself in uncomfortable situations as a reporter in order to feel that same sense of urgency that she'd encountered at Michilimackinac. Indeed, her journalism, first in New York and then in Europe, is marked by what Philip F. Gura calls a movement "beyond Emerson's cerebral and self-centered response to reform."[47] This degree of urgency coupled with Fuller's willingness to experiment with the conventions of nonfiction prose are hallmarks of the excursion form, and while Greeley's newspaper was a hotbed for reformist writing, Fuller's transition had already taken place. Indeed, it had begun as she grappled with the questions of form that underlay *Summer on the Lakes*. That sense of disturbance that she felt on the beach in 1843 stuck with her for the rest of her life. She would never be able to answer the questions that she raised that summer, but she'd found a *method* that would take her across the world.

In the fall of 1843, Thoreau, meanwhile, was living on Staten Island, working as a tutor in the home of William Emerson and struggling to make connections with influential literary figures and journalists. Fuller visited him while in the city, still charged with energy after her travels in the West. Thoreau's Journal and surviving letters do not offer a clear picture of their conversations, but Fuller surely told him about the Rock River, Michilimackinac, and Sault Sainte Marie, about the wonders and also the perils of the West. Surely Thoreau, who had published his own "Walk to Wachusett" earlier that year, was thinking about his own travel writing as he struggled to establish himself as a professional writer, and when he read *Summer on the Lakes* the next year back in Concord, he saw how to bring together his own first travel book, *A Week on the Concord and*

Merrimack Rivers.[48] Fuller provided a *method* for Thoreau's excursions. *Summer on the Lakes* is the first significant example of the excursion as an experimental, deliberately unsettling literary form, highlighting for Thoreau and others the potential of this new kind of writing. Literary writing, Fuller proved, *can* shake up readers' sense of the world, but only insofar as it considers and recreates the dramatic veers taking shape on the landscape.

2

"Some Foreign Vortex": Disturbance and the Limits of Genre

The number of French Canadian gentlemen and ladies among the passengers, and the sound of the French language, advertised us by this time, that we were being whirled toward some foreign vortex.

—Thoreau, "Yankee in Canada"

When Henry David Thoreau traveled to Quebec in September 1850, he was already an experienced travel writer. Besides "A Walk to Wachusett" (1842), among his earliest works to appear in print, he'd already made his first trip to Maine and two trips to Cape Cod, once in October 1849 and again in June 1850. He'd published his first Maine essay, "Ktaadn," in 1848 in *Sartain's Union Magazine of Literature and Art* and had lectured on Cape Cod at the Concord Lyceum and in South Danvers, Massachusetts.[1] His first book, *A Week on the Concord and Merrimack Rivers*, had been published the previous year, in 1849. Compared to his rugged first experience in Maine, the walking tours of the Cape, and the river trip that makes up *A Week on the Concord and Merrimack River*, "Yankee in Canada," should have been a simple piece recounting a short, straightforward trip north, another easy publication in a major periodical. By Thoreau's own account, however, the Canada essay would amount to something of a failure, and although partially published in 1853 as "An Excursion to Canada," it would not be printed in its entirety until after his death in 1862.[2] Today, scholars generally agree that "Yankee in Canada" is a minor essay. Still, it was a critical experiment in the limits of mainstream travel writing, and this "unmanageable" excursion was crucial in establishing Thoreau's style and vision as a travel writer.

"Yankee in Canada" emerged in tension with other, more mainstream travel writing, such as George William Curtis's *Lotus-Eating* (1852). Curtis, who had recently become associate editor of *Putnam's Monthly*, published the first three sections of "Yankee in Canada" in 1853 before a bitter dispute over certain "heresies" within the manuscript (*Corr* 2:145). *Cape Cod*, written simultaneously and ultimately among Thoreau's more mature excursions, would also be cut off when Thoreau refused to allow Curtis to remove offending material, its first four chapters appearing in 1855 and the rest remaining unpublished until after Thoreau's death in 1862.[3] In the aftermath of this situation, the contours of mainstream travel writing seemed, at least for Thoreau, to solidify, and the excursion form would take shape against a more marketable travel essay which emerged in tandem with the tourism industry. *Lotus-Eating*, though largely forgotten today, demonstrates the curious tensions emerging within the genre, including the demand that one's travel be "pleasant" and, if not repeatable, at least a model for other travels for readers back home. Thoreau's writing, meanwhile, would push beyond these constraints to depart from mainstream travel writing entirely. "Yankee in Canada," by dwelling on Thoreau's impressions and connections, demonstrates how perception of the landscape, filtered through memory and layered with significant research, shaped Thoreau's composition and revision process for years to come.

This chapter includes two sections. The first introduces "Yankee in Canada" as an experiment in which Thoreau stumbles upon an alternative sense of "America." Written at roughly the same time as *Walden* and *Cape Cod*, the Canada narrative would open the door to a decade of serious reading in non-Anglophone American history and literature, and eventually it became so intertwined with *Cape Cod* as to make the two texts ultimately inseparable.[4] "Yankee in Canada" was in the end an unsuccessful experiment, but it ultimately suggests the importance of revision and reflection for Thoreau's mature excursion form. These glimpses of Francophone America would eventually become a darker, more uncertain echo to *Walden*'s optimistic new day to dawn, which I examine in more detail in the next chapter, along with *Cape Cod*. The second section turns to Curtis's *Lotus-Eating*, which presents a stark contrast to Thoreau's commitment to "*contact*" with "the *actual* world," voiced in his 1846 essay, "Ktaadn" (*MW* 71). Curtis too wrestled with the limits of the genre, but his brand of popular travel essays ultimately embraces a market-minded escapism, turning away from the volatile conditions of the nineteenth century and the dark sensations that were coming to intrude on the literary imagination. These essays recreate and solidify one's life at home, while the excursion form attempts to disturb the reader's sense of the wider world. From these two understudied and crucially connected texts, readers can glimpse not only the emerging contours of the travel industry and mainstream travel writing, but also the methods through which Thoreau's mature excursion form would emerge.

"Yankee in Canada": An "Unmanageable" Excursion

Thoreau begins "Yankee in Canada" by gesturing toward constraints. "I fear that I have not got much to say about Canada," he writes, "not having seen much; what I got by going to Canada was a cold" (*Exc* 79). From the beginning, though, this trip proved difficult to summarize, and its themes—from religion and cultural identity to early European explorers and natural history—would defy any straightforward narrative form. Arriving in Montreal in the early afternoon on September 26 after delays in Burlington and St. Johns, Thoreau and his companion, the poet William Ellery Channing, visited the Notre Dame Basilica there, but by six o'clock, they were en route to Quebec City, taking an overnight steamer down the St. Lawrence. The next day they set out on foot, walking east along the St. Lawrence. This brief departure from the steamer schedules would come to form the heart of Thoreau's narrative of the trip, but it was still too short, too chaotic. During the next five days, Channing and Thoreau saw Quebec City, the falls at Montmorenci, the parish church at Ste. Anne du Beaupré, and the spectacular cascades on the River Ste. Anne du Nord, plus another waterfall on the Chaudière River, not to mention Montreal and Mount Royal, where he dreamed of Cartier's view from the summit in 1535. Late on the afternoon of October 1, they took the railway to St. Johns and boarded a steamboat for Burlington, returning home to Concord on the evening of October 3.

Opening lines aside, Thoreau's trip to Canada hardly amounted to "not seeing much." The trouble was that he saw *so much*, and in such a short time—from the railcars and the steamer to the cathedral in Montreal, to the walls and portals of Quebec, to the waterfalls and the mills and the villages and the intimations of the wilder, more distant reaches of Canada—that he was reeling, unable to wrestle this material into a concise, meaningful narrative. On trips to Maine and to Cape Cod, he'd traveled primarily on foot or in small boats, which allowed ample opportunity to observe the countryside. This trip, though, anticipates more characteristically modern modes of travel, in which passengers rush along in a fossil-fueled dash between "destinations," perhaps with brief, curated tastes of local culture and geography but without engaging any single environment in depth. From the start, the narrative wrestles with its abrupt changes in speed, the glaring incongruities between a steamship berth and five days walking from Quebec to St. Anne. The narrative attempts to account for the rising tourist industry, too, for the crowds of Americans and Canadians who barge into churches and clog the ferries and fill the inns and the guest houses. As he puts it in the final lines of "Yankee in Canada," Thoreau had "spent just one week in Canada and travelled eleven hundred miles" (*Exc* 163). Thoreau's volunteer literary agent Horace Greeley, the founder and editor of the *New York Tribune*, envisioned "Yankee in Canada" as a straightforward account of a tourist trip to Canada, the kind of piece that could easily find a home in New England's popular literary periodicals. The Canada essay was ultimately "unmanageable," in his words, because this short trip opened the door to something much more complicated than a tourist's account of a vacation. On foot, Thoreau's prose begins to veer, and chance encounters and associations begin to transform this essay into something different than Greeley, and perhaps Thoreau too, first imagined.

Thoreau spent all of 1851 working on "Yankee in Canada." In 1852 he delivered a lecture on the topic, but it was clear that the project was challenging his familiar ways of writing about travel and geography. He recorded long impressions of the cities, the crowds, and Catholic churches and shrines, and the essay mused on the long history of French settlement in this "new" continent. Greeley, who sensed that the manuscript was becoming quite different from Thoreau's other successful travel essays, wrote to Thoreau on March 18, 1852, urging him to break the piece up, get it in print, and move on. Still, he expressed doubt:

> As to your longer account of a canadian [*sic*] Tour, I don't know. It looks unmanageable. Can't you cut it into three or four, and omit all that relates to time? The cities are described to death; but I know you are at home with Nature, and that she rarely and slowly changes. Break this up if you can, and I will try to have it swallowed and digested. (*Corr* 2:102)

In April, Greeley wrote again, pressing Thoreau to make haste and send the revised manuscript before the upcoming tourist season began. "I fear the 'good time' is passing," he urged (*Corr* 2:281).[5]

While Greeley attempted to get the Canada narrative "swallowed and digested," Thoreau was trying to find a home for the early chapters of *Cape Cod*, which seemed more promising. He sent the *Cape Cod* manuscript to George William Curtis, editor at *Putnam's*. Curtis agreed to publish the piece in installments, printing it anonymously, like all the other contributions. Greeley, meanwhile, unaware, convinced the magazine's owner, George Palmer Putnam, to publish "Yankee in Canada." The miscommunication grated on Thoreau, and it would ultimately leave both works in flux for the rest of his life.[6] Worse yet, when he received the January 1853 number of *Putnam's*, which included the first installment of "Yankee in Canada," he found that Curtis had cut a sardonic line: "I am not sure that this Catholic religion would be an admirable one if the priest were quite omitted. I think that I might go to church myself sometimes, some Monday, if I lived in a city where there was such a one to go" (*Exc* 88–9). Thoreau threatened to withdraw the manuscript, and according to Joseph Moldenhauer, he may have gone so far as to demanded that the manuscript be returned immediately, though publication would continue, at least for the time being.[7] A second installment of the essay appeared in the February 1853 issue, ending with a parenthetical "To be continued." In March, a third installment was published, including more than half of the "St. Anne" section, but then discussions with Curtis apparently broke down.

In a letter to his friend H. G. O. Blake, who'd been reading the piece in *Putnam's*, Thoreau writes:

> I do not wonder that you do not like my Canada story. It concerns me but little, and probably is not worth the time it took to tell it. Yet I had absolutely no design whatever in my mind, but simply to report what I saw. I have inserted all of myself that was implicated or made the excursion. It has come to an end at any rate, they will print no more, but return me my mss. when it is little more than half done, as well as any other I had sent them, because the editor Curtis requires the liberty to omit the heresies without consulting me—a privelege [*sic*] California is not rich enough to bid for. (*Corr* 2:145)

Thoreau's admission here, that he had "no design whatsoever in mind," suggests that his prose is a straightforward, linear account of a journey, but this comment belies a much more strategic mode of narration and reflection on the history of North America. None of this, however, would appear in print during Thoreau's lifetime.

Scholars have generally described this editorial disagreement and resulting suspension of "Yankee in Canada" in terms of Thoreau's personality, but this episode also suggests that Thoreau was actively reimagining the purpose of his work. For Thoreau, pieces like the Canada essay should, as he put it to Blake, record all of oneself "that was implicated or made the excursion." Such writing is not a model for some repeatable route or vacation, but a record of one's evolving sense of self and world. Further, in encountering such a text, the reader should be challenged or even disturbed, should confront some aspect of the trip that unsettles or surprises or simply defies expectations. These "heresies" are not merely a hallmark of Thoreau's character and beliefs about writing, but a signature element of his emerging excursion form. As in his puns and the other slippages of meaning that characterize so much of *Walden*, his travel writing is meant to quite literally shake up our preconceived sense of the world, the text, and language itself, to wake us up.

What limited attention "Yankee in Canada" has garnered has largely reflected those dismissive remarks in 1853, when the essay was being pulled from *Putnam's*. Robert Richardson writes that for the most part, "Thoreau's account of the trip is thin, uninspired, complaining."[8] Similarly, John Aldrich Christie notes that "Yankee in Canada" "lacked, except in occasional spots, the immediacy of Thoreau's keen observations and involvement that gave vitality to his later accounts of Maine and Cape Cod."[9] Walter Harding even goes so far to call the essay "one of Thoreau's least inspired 'excursions'" with "atypical, staccato" sentences that amount to "pedestrian journalese."[10] Despite this prevailing sense of the narrative's shortcomings, some scholars have taken the Canada essay seriously, especially for its political and social commentary. Joseph Basile Lawrence draws from the February 1853 letter to Blake to argue that despite Thoreau's own frustration with the essay and the pace of the trip itself, "Yankee in Canada" is significant for its reflections on the power of social, national, and religious institutions, particularly in so far as they limit the kind of conscious progressivism that Thoreau so values in American culture.[11] Stephen Adams takes this argument further, suggesting that "Yankee in Canada" presents the reader with a series of images which, although seemingly hasty, random, and disjointed, ultimately develop into a "network of associations" that suggest a cohesive structure, with an attempt at "an honest walk" in the center as a climax.[12] For Adams, the essay is not so much a travel essay, successful or otherwise, but an "account of a failure of vision" and an "exploration of the social forces behind that failure."[13] More recent scholarship by Patrick Lacroix has continued in this vein. As he puts it, the essay "reveals great faith in the promise of republican values epitomized by the American Revolution" while still resisting "blind patriotism."[14]

Together, this scholarship suggests that while the Canada essay is not perhaps successful in the conventional sense, it is nonetheless significant with regard to Thoreau's later excursions, especially the travel essays which explicitly take up themes of national origins, previous exploration narratives, and the ultimate *purpose* of travel writing, that disturbance to a preconceived vision of the world. I share Adams's sense that "Yankee in Canada" is, in fact, a structurally cohesive account of a certain kind of failure, that it reflects a movement toward something new altogether. If other accounts of Quebec and Montreal aim to highlight the novelty of a French Canadian vacation for readers back home, Thoreau would instead offer a glimpse of something that profoundly unsettled his sense of Americanness, but this disturbance to his established way of being in and writing of the world is only just beginning to influence the final shape of the text. Thoreau's prose thus emphasizes a sense of uncertainty and curiosity.

Thoreau would begin to puzzle over his experiences in Canada almost as soon as he returned home. As Laura Dassow Walls puts it, "he plunged into a new course of research, sketching a kind of sweeping, cosmopolitan, *longue durèe* history … of Greater New England conceived across deep time and planetary space."[15] Thoreau had not, that is, really *seen* Canada at all, but he had gotten a taste—and that would be just enough to set him off. In other words, we might read "Yankee in Canada" as, in Greeley's words, an "unmanageable" essay and an affective hinge, or a project that, because it didn't sit well with Thoreau's sense of the world, set off a cascade of reading and further writing, an attempt to reimagine just what it is that a travel essay can offer, can do for both the author and the reader. "Yankee in Canada" thus anticipates a more unsettled, more disturbing sense of travel writing. It is, moreover, the necessary precursor to his more radical, disturbing excursions, such as *Cape Cod* and *The Maine Woods*.

"Yankee in Canada" focuses on an emerging affect, a sense of something that is not yet definite and that can only be represented by a vaguely conscious notion of what might come to be. I call this indeterminate sense of something "affective resonance," a term that captures the ways that certain sensations can produce a vibrant, almost enchanting allure even as we only partially or glancingly encounter them. If Thoreau's life in Concord represents a kind of sticking point, a tension that emerges as he cannot *quite* fit into the spirit of Yankee industriousness and civic republicanism, then this trip to Canada becomes a moment of dramatic awakening that will linger with him for an extended period of time, a feeling that he cannot quite identify but that he cannot shake either. He has a good sense of the French language, but the Quebecois pronunciation baffles him. The landscape is similar to what he's seen in northern Maine and even Massachusetts, and yet it seems a world apart. The frustration that we find in the first chapter of *Walden*, "Economy," forms a silent backdrop to Thoreau's trip to Quebec. How ought one live in the world? How, in the face of the rising US industrial capitalism, can one find or create the space to live deeply and purposefully in a rapidly changing political, cultural, and ecological moment? What would such a project look like under the watchful eyes of the Catholic church and the Canadian government?

Thoreau went to Canada with a certain sense of how to address these predicaments, but French Canada would challenge those approaches in profound ways. As Lacroix notes, "Yankee in Canada" reflects a moment when the political reality that Thoreau knew in Concord, a brand of civic republicanism built on thoughtful, engaged individualism, was being fundamentally challenged. That is, a social, cultural, and ecological world was ending with the dawn of a new political and economic moment, but even then, alternative possibilities were available.[16] Thoreau was aware of those changes, and he saw them in the massive environmental, social, and cultural reorganization that took place following the rapid rise of the railroad in rural New England farming villages.[17] French Canada, even though it stymied him, offered certain resources for reconsidering the looming changes over the New England landscape, and describing his trip there would necessitate a series of veers—both literal, on a forest path, and literary, in the types of prose and the connections engendered in Thoreau's reflections. By attending to these veers as part of a process of revision and reflection, a process somewhat harried by the literary marketplace, we can begin to see through that opening line to glimpse both the structure and, more importantly, the encounters and ideas that resonate.

Thoreau's jokes aside, the text does suggest a lasting interest in French Canada and in a narrative that might capture his impressions of the place. "Yankee in Canada" pushes back against mainstream expectations in several ways, beginning with the pace of the journey. By focusing on the purposefully slow, deeply intentional engagements with place in this excursion, Thoreau resists the rising tide of industrial travel and tourism, insisting that we drink deeply of our experiences beyond home, turning over those encounters and affects to challenge mainstream commodification of experience and the compression of distance via modern travel—whether by train, carriage, and steamer or by plane and automobile. "Yankee in Canada" furthermore suggests that when challenged, when unsure of our experiences and their value, we use those sensations to somehow see things a little differently. This ultimately becomes the design of the essay—to give these veers their form within the overall pace and structure of the narrative, that they might allow readers, and perhaps Thoreau himself, to better see what *sticks* after this excursion to French Canada.

Consider, for instance, the third chapter, "St. Anne," which describes a small village outside of Quebec City and recounts Thoreau's detour to an impressive waterfall in a rugged canyon just beyond, the climax of his ten-day excursion and the limits of his walk along the St. Lawrence. Thoreau begins the chapter by emphasizing the geographical particulars of the place—thirty miles from Quebec, he is as far as from home as this trip will take him, and as close as possible to an authentically Canadian experience. He and Channing are, as Thoreau puts it, "in that part of Canada which was first to be settled, and where the face of the country and the population have undergone the least change from the beginning, where the influence of the States and of Europe is least felt, and the inhabitants see little or nothing of the world over the walls of Quebec" (*Exc* 111). Walking through the village, they muse on the region's agriculture and the history of its settlements, marveling at what is a thoroughly foreign culture in a landscape strikingly similar to their native New England.[18] The village of St. Anne, as Thoreau recounts it, is not so far removed from Concord, but it seems a world apart. "There were no guide-boards because there was but one road," Thoreau writes, "there were no shops nor signs, because there were no artisans to speak of, and the people raised their own provisions; and there were no taverns because there were no travelers" (*Exc* 119). When they inquire about lodging, they are directed to a nearby house, where they spend the night and take breakfast the following day.

The next morning, a Sunday, Thoreau and Channing depart the village shortly before mass at the Church of La Bonne Ste. Anne, passing the worshippers on the road as they head for the falls. But as in many of his later excursions, Thoreau quickly loses his way. Being lost will later prove essential to Thoreau's mature excursion form. In this moment, however, being a little disoriented is merely a thrilling moment along the way to a notable landmark:

> We followed for a short distance a foot-path up the east bank of this river [the St. Anne], through handsome sugar-maple and arbor-vitae groves. Having lost the path which led to a house where we were to get further directions, we dashed at once into the woods, steering by guess and by compass, climbing directly through the woods, a steep hill, or mountain, five or six hundred feet high, which was, in fact, only the bank of the St. Lawrence. Beyond this we by good luck fell into another path, and failing this or a branch of it, at our discretion, through a forest consisting of large white pines,—the first we had seen on our walk,—we at length heard the roar of falling water, and came out at the head of the Falls of St. Anne. (*Exc* 122)

This is the most direct, significant example of the veer in "Yankee in Canada," of the change in direction and pace that Thoreau would make central to his later travel writing. Having slowed down, we deliberately and earnestly depart from those parishioners, from the churches, from the essay's rushed accounts of cities and the schedules of steamers and trains and ferries. Here, in the woods, a chance encounter leads to a surge of new associations. Encountering a grove of white pines, Thoreau seems poised to consider the similarities between this forest and the woods of Massachusetts, of Maine. Then, with this in mind, and with his thoughts racing from the twists and turns and chance forks in their route, he approaches the falls at St. Anne, where his characteristically detailed, even *over*-detailed description seems to fade, the narrative slowing and changing direction. The narrator, thinking about the river more broadly, is spellbound to the point that he cannot offer an objective measure of the height of the falling water:

> We had descended into a ravine or cleft in the mountain, whose walls rose still a hundred feet above us, though we were near its top, and we now stood on a very rocky shore, where the water has lately flowed a dozen feet higher, as appeared by the stones and drift-wood, and large birches twisted and splintered as a farmer twists a withe. Here the river, one or two hundred feet wide, came flowing rapidly over a rocky bed out of that interesting wilderness which stretches toward Hudson's Bay and Davis's Straits. Ha-ha Bay, on the Saugenay, was about one hundred miles north of where we stood. Looking on the map, I find that the first country on the north which bears a name, is that part of Rupert's Land called East Main. This river, called after the holy Anne, flowing from such a direction, here tumbled over a precipice, at present by three channels, how far down I do not know, but far enough for our purposes, and to as good a distance as if twice as far. It matters little whether you call it one, or two, or three hundred feet; at any rate, it was a sufficient water-privilege for us. (*Exc* 122)

This passage demonstrates the value of that literal and narrative veer within the overall structure of the narrative. In this vague account of the height of the falls, it becomes clear that what Thoreau takes home is not so much a sense of the waterfall itself as of the course of the river through time and in space, a route for further explorations, whether real, imagined, or both, of the Canadian interior, not so much repeatable as phenomenological. The actual height, some one hundred or two hundred or three hundred feet, is less important than the headwaters in the "interesting wilderness," somewhere up there in the fantastically wild expanse of the Canadian interior. And rather than its size, Thoreau seems more impressed by the canyon's totality, its "most wild and rugged and stupendous chasm, so deep and narrow where a river has worn itself a passage through a mountain of rock" (*Exc* 124). Instead of recreating the site of one's travel, this passage shifts our focus to the unrepeatable nature of experience, to the waterfall in that moment, as Thoreau and Channing first encountered it after their wandering course through the forest. In doing so, the text comes to register not merely the locations seen, but also the senses and impressions and connections that it inspired. Here, the entire trip seems to change, becoming a portal to a wider view of the North American continent.

This moment along the St. Anne is remarkable primarily because of this cascade of connections, which draw from the initial moment of encounter to suggest relations to other times and places, to things that Thoreau has read, experienced, or merely dreamed of. These connections are best explained in terms of affective resonance, or the *sticky* sense of something interesting *out there*, a potential "somewhere" on the horizon that is suddenly thrown into relief by one's impression of something else, near at hand. Affect is here a kind of spatiotemporal connection or, in Sara Ahmed's words, "what sticks, or what sustains or preserves the connection between ideas, values, and objects," between the falls and that intoxicating sense of what Canada, what all of North America, might well look like."[19] And the particular affect that Thoreau draws from in "Yankee in Canada" is even more *sticky* than this imagined map might first suggest, because he also recognizes that this region has a much longer history of Euro-American settlement than his native New England. After his trip, while writing from his home in Concord, this short stretch of one particular river comes to resonate not merely with his one glance at a map, but with all of his previous thoughts, experiences, and associations—with books on Canada and the Arctic, with the fur trade, with canoes and rocky rivers and waterfalls wherever he might encounter them. Some of these connections are immediate and vivid, while others are more subdued. From this moment, though, Thoreau will see this canyon as one point, one part of a continent situated in time and space, in a continuum from the local to the continental.[20] Here, on the banks of the Ste. Anne du Nord, Thoreau encounters a waterfall as a portal, a point from which he can access a larger world that stretches far beyond his normal bounds. From this point on, he can write, or rather *rewrite*, the North American continent not just from Concord, but from a web of glancing encounters, partial connections, and imaginative associations that stretch throughout the northeastern reaches of the continent. In this way, one veer engenders more, and connections between texts and ideas and experiences began to converge and coalesce.

This moment is also significant because it demonstrates the importance of pace within Thoreau's emergent excursion form. Ultimately, this narrative downshift suggests, as Timothy Clark puts it, that "the speed with which the world happens for a human being need not be the norm."[21] As in animal narratives, time-lapse photography, slow-motion cinema, and similar forms, travel narratives can disrupt normal manners of seeing and moving through the world and thus "explore unknown modes of experiencing time and space."[22] Here we see that the speed of human life can change, and in the process new encounters and new sensations become suddenly possible. New insights are suddenly thrown into relief. "Yankee in Canada," even as it struggles to capture the incongruities of this trip, comes to suggest the value of veers away from modern conceptions of travel, from fast-paced, fossil-fueled modes of navigating and living in the world. In his account of a trip to Canada, Thoreau would allow space and time for meditations on history and culture, for comments on the separation of church and state, for observations on language, and for a discussion of Canadian geography. First, however, his prose must enact that crucial change of pace and direction.

St. Anne was, as Thoreau puts it, "the limit of our walk down the St. Lawrence" (*Exc* 124). As he leaves the falls, he muses on the French language and its uniquely poetic depiction of the landscape before him, its uncanny compatibility with the St. Lawrence Valley and its rocky rivers. French Canada has clearly left a mark on Thoreau even as it has unsettled him. The trouble, and the reason, perhaps, that the essay will be so "unmanageable," is that Thoreau's response is deeply conflicted. The wild expanse of country beyond Quebec has thrilled and amazed him, but he cannot quite square the place with his sense of Yankee New England and this "new" continent. This river, and the St. Lawrence as well, might lead him deep into the wild heart of the North American continent, but it is also lined with some of the oldest European settlements in North America. He describes an acute sense of conflict:

> To a traveller from the Old World, Canada East may appear like a new country, and its inhabitants like colonists, but to me, coming from New England, and being a very green traveller withal—notwithstanding what I have said about Hudson's Bay,—it appeared as old as Normandy itself, and realized much that I had heard of Europe and the Middle Ages. Even the names of humble Canadian village, affected me as if they had been those of the renowned cities of antiquity. To be told by a habitan, when I asked the name of a village in sight, that it is *St. Fereole* or *St. Anne*, the *Guardian Angel* or the *Holy Joseph's*, or of a mountain,

> that it was *Belange*, or *St. Hyacinth's!* As soon as you leave the States, these saintly names begin. (*Exc* 125)

One would expect Thoreau to revel in this sense of a wider world, so richly detailed and evocative, but in fact he is puzzled. He goes on:

> I began to dream of Provence and Troubadours, and of places and things which have no existence on the earth. They veiled the Indian and the primitive forest, and the woods toward Hudson's Bay were only as the forests of France and Germany. I could not at once bring myself to believe that the inhabitants who pronounced daily those beautiful, and to me, significant names, lead as prosaic lives as we of New England. (*Exc* 126)

In his response to this impression of the place and to the French language, we see Thoreau's process of working through his experiences in Canada, of trying to settle his impressions with his more "prosaic" life in nineteenth-century New England. This is the aftermath of that *sticky* affect—the troubling awareness of a world split open and immediately wider, of ontologies that are unsettled by an enormous and sudden broadening of referents and contexts. In his trip to Canada, Thoreau's Concord is blown open, and he must come to terms with an America that is no longer New England-centric, that does not grow out of the Puritan settlement at Plymouth, but from a decidedly French (and, as Thoreau writes elsewhere, Spanish, Dutch, and even Norse) history of North America. As Moldenhauer details, Thoreau continued working on "Yankee in Canada" after his dispute with Curtis in 1853, but the text remains largely unfinished because of the depth and scope of this research, and because his attention drifted to other works, including *Cape Cod* and *Walden*. If Thoreau struggled to finish the essay, to tie it all together in some comprehensive way, it is because it is not, as Greeley imagined it, a simple travel essay, nor is it, as Thoreau himself said, entirely free of design. "Yankee in Canada" refuses to recreate a vacation, to trace a route toward some remote waterfall, even to isolate the five-day walk from the rest of the trip. Thoreau would use it all, and rather than recreating a tour of Canada, he would open a portal to a new sense of the continent. *This*, he suggests, is the purpose of an excursion, and this will be the standard to which he holds all travel narratives written afterward.

In the wake of his trip to Canada, Thoreau began to research the historical contingencies that led to the social, political, and ecological changes that he was already charting in Concord, and much of the material that he begins to explore in "Yankee in Canada," including extensive research on French explorers and cartographers, made its way into the final chapter of *Cape Cod.* John Hessler, writing on Thoreau's Indian and Canadian notebooks and the maps that he traced from Champlain's *Voyages,* writes of this as a shift away from Emersonian Transcendentalism toward the geographically oriented sciences of Darwin and Humboldt, a move that allows us to access what was, in fact, "a geographical link between natural and civil history."[23] In this blossoming interest in French exploration and history, Thoreau swerves away from Greeley's pragmatic sense of travel writing toward something else, something not yet determined within this single essay. Without such a turn, he might never have been able to suggest that a traveler stand on Cape Cod and "put all America behind him" (*CC* 215).

George William Curtis and the Literary Marketplace

While Thoreau began drafting the Canada essay, his editor at *Putnam's* was putting the finishing touches on his own travel book, which would come to represent another means of figuring disturbances within a literary travel narrative. George William Curtis's *Lotus-Eating: A Summer Book* was published by Harper & Brothers in the late summer of 1852, a collection of eleven chapters that had first appeared over the summer of 1851 as travel letters in Greeley's *New York Tribune.* Printed alongside spare, serene illustrations by the Hudson River School artist John Frederick Kensett, Curtis's essays recount a six-week trip across New York and New England, from the Catskills and Niagara to Lake George, Nahant, and Newport.[24] Like other travel narratives published in the periodicals of the day, these essays draw from Romantic landscape aesthetics to describe and critique the emergence of the tourist industry in the northeastern United States, with accounts of steamship and stagecoach travel, of old and new resort communities, and of popular natural attractions. Along with his two previous books, *Nile Notes of a Howadji* (1851) and *The Howadji in Syria* (1852), both recalling a four-year tour of Europe, Egypt, Palestine, and Syria, *Lotus-Eating* would help establish Curtis as a travel writer on the international stage. Ultimately, *Lotus-Eating* reflects the changes that were taking place in the US literary marketplace, the period during which travel writing came to reflect first the imagination of the individual traveler and eventually the emergence of a tourist industry.[25] As a result, the book's lyrical, associative essays ultimately leave the reader, like Curtis himself, unable to account for the American scenes before him.

Today Curtis is a relatively minor figure in American literary history, but during the nineteenth century he was well known. In 1842 he enrolled as a "boarder" at Brook Farm in West Roxbury, Massachusetts. Although best known as a radical experiment in communal living, Brook Farm also offered a six-year course of education for college-bound men, and Curtis, along with his brother Burrill, spent eighteen months there, studying under a variety of influential teachers and participating in the regular masquerades, balls, and concerts. He also became acquainted with many of the regular members and the other boarders, as well as the regular visitors, including Bronson Alcott, Margaret Fuller, and Emerson. Ultimately, though, he would leave Brook Farm, writing that "a wiser man lends himself to no organization."[26] In October 1843, the Curtis brothers returned to New York before moving to Concord to continue their education, striking a deal whereby they spent the mornings working on the farm of Captain Nathan Barrett in exchange for room and board, leaving their afternoons free for exploring, studying, and visiting with Emerson, Hawthorne, and occasionally Thoreau. In 1844 he joined Emerson's short-lived effort to establish a club of like-minded Transcendentalists, the group meeting three times before breaking apart, and in 1845 he helped Thoreau build his house at Walden Pond.[27] This period at Brook Farm and in Concord would leave a permanent impression on Curtis, but despite his affinity for Transcendentalism and his relationships with so many mid-nineteenth-century literary figures, his own literary career owed more to an identity as the "Howadji," or traveler.

In August 1846, Curtis sailed from New York, destined for Marseilles. He spent the next four years touring Europe, with detours to Africa and the Near East, and when he returned to the United States in the late summer of 1850, he found himself with a promising body of literary material. Almost immediately he set about turning his notes into *Nile Notes of a Howadji*, which was published in March 1851. The book was reasonably successful, and though reviews weren't uniformly positive, it led him into magazine writing almost immediately. He had already been publishing reviews and critical commentaries, and that summer he accepted an assignment as a travel writer with the *New York Tribune*, agreeing to tour a series of old and new resorts throughout the Northeast. He journeyed from New York up the Hudson to the Catskills, going as far west as Niagara and north to the Adirondacks before returning to the coast in August. Dispatching his final letter in September, he set to work finishing *The Howadji in Syria*, which would appear in April 1852, also based on his notes from those four years abroad. Then he began revising his American travel letters from the *Tribune*, which would be published that summer as *Lotus-Eating*. In just two years, Curtis had established himself as a prolific and successful author, and in the fall, he began work as an editor at the promising new periodical, *Putnam's Monthly*.

In all his travel books, Curtis drew from a cosmopolitan point of view developed while traveling in Europe, crafting a unique landscape aesthetics and a comparative approach to art and culture. He was by no means the only American author to consider European landscapes and cultures. Other books from the period offered a similar sense that Americans had much to learn by looking toward the Old World. Many Transcendentalists and like-minded writers produced works in this vein, including James Freeman's Clarke's *Eleven Weeks in Europe* (1852), C. A. Bartol's *Pictures of Europe* (1855), Emerson's *English Traits* (1856), and Fuller's posthumous *At Home and Abroad* (1856), which collected her dispatches from Europe along with *Summer on the Lakes*, appearing six years after her death. Lawrence Buell notes that these Transcendentalists realized that they were traveling on well-worn ground, as it were. Unlike frontier narratives, which emphasized the originality of specific experiences and locations, these accounts of Europe would need to offer new points of view. Their "contribution" or success would thus rely on "deploying the resources of their wit, descriptive ability, capacity for original reflection, and eye for out-of-the-way detail."[28] By looking toward Europe, these books also helped place Transcendentalism on a larger, international stage, and though they made only minor ripples in the US literary imagination, they did contribute to a growing field of travel literature that offered more than simple reportage of distant places. Transcendentalist travel narratives also helped establish a critical trend whereby later travel writing would be judged partly on the merit of its description, but also on the author's personal reflections, rather than simply the novelty of the places being visited.

Curtis, though initially influenced by Emersonian individualism, seems to have departed from Transcendentalism when he set out for Europe. His travels from 1846 to 1850 were more a Wanderjahr than a Transcendentalist's trip abroad, and his first two books reflect his growing cosmopolitanism and a sense for American readers' interest in the wider world. Outside of the Transcendentalist circle, many American authors were capitalizing on this interest. Herman Melville had gained his first literary fame with *Typee* (1846), a fictionalized account of adventures in the South Seas. Beginning around the same time, Bayard Taylor would use an advance for travel articles to kick off a decades-long career as a poet, travel writer, and diplomat. His second book, *Views*

Afoot, or Europe seen with Knapsack and Staff (1846), collected these dispatches in two volumes and led to further assignments in California, Africa, Palestine, and the Mediterranean, and eventually to India, China, and Japan.[29] Curtis's own traveling companion, George S. Hillard, wrote *Six Weeks in Italy* (1853), and William Wells Brown produced two accounts of his travels in Europe, *Three Years in Europe* (1852) and *American Fugitive in Europe* (1855). By 1850, travel writing had become an obvious route for Americans to establish themselves as writers on the national and international stage. When Curtis returned from Europe, he was at the beginning of a wave of popular travel writing that would compare European landscapes, culture, and cosmopolitanism to the echoing calls for an American literary nationalism. Mark Twain, William Dean Howells, Henry James, and others would take up the same project later.

In his first two travel books, Curtis offers his readers a sense of an American observer of the larger world, of the national character distilled into the roaming observations of this singular "Howadji." Even *Lotus-Eating*, which purports to focus on American resorts and tourist sites, draws from this global perspective, each chapter looking back on Curtis's experiences in Germany, Switzerland, France, Italy, and beyond. In the first chapter, "The Hudson and the Rhine," Curtis describes the pleasures of a steamship voyage up the Hudson, and unlike the dissonance and bustle that so grate on Thoreau in Quebec, Curtis claims to prefer traveling this way. "It is only in the steamer that the Hudson can be truly perceived and enjoyed," he writes, offering "silence, cleanliness, sufficient speed, and an unencumbered enjoyment of the landscape." The railroad, by comparison, offers "a swiftly unrolling panorama of dreams," but yet "would have roused Rip Van Winkle himself, and the dust would have choked and blinded him as he opened his eyes" (*Lot* 13). For Curtis, there is no option but to be whisked along by rail or by steamship, and although it seems preferable that the thoughtful, poetical American to travel by water, the character of Irving's New York landscapes ultimately still proves inaccessible.

If Thoreau's Canada essay suggests that an excursion is unrepeatable, that the form of a travel essay must reflect the dynamism of specific places, indeed, of the traveler himself, Curtis draws from the relative stability of well-known tourist sites and literary archives to suggest a degree of permanence to his routes, the places he visits, and the experiences that he describes. It is "curious," Curtis writes, that he shares the deck with no English tourists, whose "sublime skepticism of any excellence out of England is pleasanter than our [American] crude boastfulness and subserviency" (*Lot* 14). The delights of the Rhine, he goes on, "are those of romantic association and suggestion. They are those which are only possible in an old and storied country." He continues:

> I know that romance is in the poet's heart, and not in the outward forms he sees. But there is the technical material of romance—the moonlight, a ruin, an Italian girl, for instance—which is useful in begetting a romantic mood of mind, as a quotation which will suggest verses that haunt you all day long. And it is in this material that the Rhine is so rich. (*Lot* 22)

Standing on the deck of this steamer, Curtis seems entirely removed from the landscapes and literary nationalism of Irving, a product of this decidedly new period of American literature. He thus not only compares the Hudson to the Rhine, but finds the American landscape lacking this material, and therefore the poetic suggestions he is looking for.

By embracing tourism and a more global perspective on travel, Curtis also ultimately reframes earlier conceptions of the American landscape as a "wilderness" or a "frontier." James Weaver, one of the only scholars to write on Curtis in recent decades, argues that these glances back toward the Old World reflect the tensions and contradictions between Curtis's experiences of the nascent American tourist industry and his yearning for a "lost connection to Europe."[30] As Weaver puts it, "*Lotus-Eating* provides a distinct vision of the emergence of nature as a recreational commodity and the transformation of wilderness into a surrogate home for the traveler."[31] This sense of "nature" as a destination, suddenly accessible and welcoming because of an industry of mechanized travel and resort communities, would continue to gain traction in the latter half of the nineteenth century. In 1858, for instance, Emerson, along with James Russell Lowell, Louis Agassiz, Jeffries Wyman, William James Stillman, and five other intellectuals made a trip to Follensby Pond in the Adirondacks, establishing a month-long residence in what came to be called the Philosophers' Camp.[32] Although inspired by a romantic sense that the region's natural features would lead to important developments in American culture, the group's time there was in fact at the beginning of a wave of popular travel. Following the Civil War, the guidebook *Adventures in the Wilderness; or, Camp-Life in the Adirondacks* (1869), written by the Boston clergyman and lecturer William Henry Harrison Murray, so popularized the region that the hordes of tourists there would come to be called "Murray's fools."

Even as he anticipates and celebrates the emergent possibilities of an American tourism industry, Curtis offers an uncommonly harsh critique of the landscapes that he passes through. “One would be loth [*sic*] to exhort a European to visit America for other reasons than social and political observation, or buffalo hunting,” he writes. Part of his critique is simply a matter of scale, of New England’s comparatively smaller mountains, less developed transportation networks, and smaller cities. At the heart of his critique, however, is a belief that objects and landscapes are interesting or beautiful primarily in terms of their associations. “The *idea* of the great western rivers and of lakes as shoreless to the eye as the sea, or of a magnificent monotony of grass or forest, is as impressive and much less wearisome than the actual sight of them,” he concludes (*Lot* 138–9). Curtis thus ultimately records and celebrates the grandeur of American nature, the potential for these places to inspire future travelers, even as he dismisses them for the contemporary, cosmopolitan traveler. Further, as Weaver writes, “he remains alienated from the American scene he tours, primarily because it fails to match his earlier experiences of Europe.”[33] In comparing the Hudson to the Rhine and attempting to understand his American journeys in terms of his time in Europe, Curtis enacts a nostalgic return to that former period of his life. In searching for the “material of romance,” his travel writing mourns the ephemerality of experience, the distance between the present and a period of one’s life that is now irrecoverable. This sense of melancholy ultimately lends a strangely anti-nationalist bent to *Lotus-Eating*, an unsettling sense that the American landscape is not enough to break free from the influence of Europe, that the American travel writer in Europe will become, in fact, something of a European, and that the American traveler in America will always yearn for something more, something new. Moreover, Curtis emphasizes the repeatability of his travels even as he casts his gaze backward to previous experiences.

Lotus-Eating stands out among Antebellum US travelogues. It is neither an account of the distant West nor of some foreign experience, of interest to nineteenth-century readers primarily because of its remove from the daily life of those in New England and New York. Unlike Thoreau’s excursions to Canada and Cape Cod, it is not oriented toward home, and it offers the reader only very curated senses of American scenes. Rather, Curtis suggests that his various destinations are significant in large part because of their proximity to urban centers, their accessibility and amenities for vacationers. The Hudson, for instance, however “larger and grander” than the Rhine, “is not to be devoured in detail,” though it is unlike any other river in Europe or elsewhere (*Lot* 22). “If a man would touch the extremes of experience in a single day,” Curtis writes, “I know not how it could be better done, than by stepping upon a steamer, after a long bustling morning in Wall-street, and reading the 'Culprit Fay' by moonlight upon the piazza of the hotel at West Point, looking up the river to Cro' Nest" (*Lot* 24).[34] The Hudson, he suggests, is most interesting in terms of its contrasts, in its proximity to both the center of New York cultural and economic life *and* its associations with a still-developing American culture, a "continent behind," a place of forests rather than vineyards, of "imposing mountains" rather than "grain-goldened heights" (*Lot* 22–3). The Hudson thus makes for an interesting vacation from New York, a pleasant midsummer escape, but if "the wild grape in our woods" suggests distant landscapes and even inspires some poetry of its own, it is merely from "the association with Rhenish vineyards, and they in turn from their association with the glory of the grape in all literature and tradition" (*Lot* 21–2). The "continent behind" here is both an interesting American interior, beyond the coast and the Appalachians, and, at the same time, a culture that is less developed, less *interesting* than Europe.

Curtis will ultimately confront the ways in which travel can leave one shaken and uncertain, but unlike Thoreau in "Yankee in Canada," *Lotus-Eating* will maintain the vacation as its final form. Moreover, it will consider travel in a relatively shallow sense. On the Atlantic coast, for instance, the book can take stock of the changing tides of fashion, but the beaches, the changing moods of the ocean pose a challenge, even a threat. In the final essay of the book, "Newport," Curtis describes the fogs that arise in early summer and autumn, those "not always pleasant" moments when the moods of the ocean seem "to brood over your spirits with a dull gloom." These fogs come and go sometimes, approaching and then retreating over the course of the day, visible on the horizon or up the shore. Then in the evening, it "beleaguers" and "overcomes" one's immediate surroundings; it "penetrates at windows and doors." A band might play, a party might go on, but this "ghost of the ocean" "asserts the sea's sovereignty over land" (*Lot* 196). For a moment, all the "gayety of Newport is suddenly quenched" and the outside world, so dark, gloomy, uncertain, presses in on the scene (*Lot* 197). Curtis's pleasant escape, the reader senses, is threatened by this reality of life in the Atlantic world. Any less pleasant affect, the reader comes to sense, anything which disrupts the *idea* of a pleasant summer retreat from the city, is to be avoided.

In the final pages of *Lotus-Eating*, Curtis seems poised to confront this darker mood, this atmosphere which would interrupt one's evening, even one's entire vacation. The book opened with a paean to midsummer, to the beauty of an afternoon's escape from the city, even reflecting on the insufficiency of that word, that idea, *beauty.* Here, though, as early autumn sets in, Curtis's reverie is different. Riding out in a heavy fog, "the chaotic sound of the sea" rising out of the gray, he dreams of Coleridge and Turner, each standing alone and looking out over London, obscured by the smoke and vapor of rising modernity. What did Coleridge *see* in those low, thick, noxious clouds? There *must* have been "more than the image of his own mind and his own career" within "that limitless city, wide-sparkling, many-turreted, fading and mingling in the shining air," perhaps "strange voices calling from its clouds" (*Lot* 199). And for Turner, London must have been even more a marvel. "There is no grander lesson in the history of any art," Curtis writes, than Turner's sense of that mist, "which to others was but a clog and inconvenience" (*Lot* 199). How, Curtis seems poised to ask, can we come to really see this world right before us through all of the smoke and the haze? How can we come to see the very fog itself?

That evening Curtis finds himself alone on the beach, "surrounded by fog" (*Lot* 199). He describes the ocean pressing in yet somehow remaining distant, everything obscured save the moon, which itself comes and goes through gaps in the clouds. It is as though he is walking in a strange, unsettling dreamland:

> I seemed to be upon the bottom of the sea, for nothing was visible save occasionally the moon, as the fog thinned over my head—the seemingly circular spot of beach upon which I stood—and the long, white seething line of surf that fell exhausted along the shore. The confused moan of the sea was the only and constant sound. Fascinated by the strangeness of the scene, lost in the fog, whose murky chill lay damp upon my hands and face, I wandered over the beach. I ran, but could not escape the small round spot of black beach—the encompassing dead white cloud—the moon, blotted out again and revealed. I shouted aloud, but my voice fell flat and lost, and the murmur of the surf boomed in melancholy mockery. I stood still, but the continuous sound did not destroy the weird silence. I ran to the edge of the sea; the water broke over my feet and slid far up the beach and washed my tracks away. (*Lot* 199–200)

In this moment, the only like it in *Lotus-Eating*, Curtis is unable to describe his experiences in terms of the picturesque. This is not quite the sublime, but rather an overwhelming sense of uncertainty and chaos, the "confused" sound of the sea overtaking his own voice, "flat and lost," and washing away his tracks. Without the opportunity to look backward, to take in the sweep of the beach and the horizon, Curtis struggles to describe the scene before him. Disoriented, he panics, but whether running or standing still, he is confronted by the same sense of directionlessness, the same sense of being trapped. The ocean seems finally to have swallowed him, to have left this pleasant sketch of seaside life inundated, lost, subsumed by a suffocating sense of uncertainty.

Wandering further amidst this fog bank, Curtis becomes even more disoriented. As he walks, shapes begin to emerge from the vapors:

> I advanced constantly with no sense of progress and saw suddenly a huge, fantastic figure looming ominously through the fog-cloud and confronting me. I stopped as if an army had risen before me, then ran toward the figure which dwindled into a shapeless block, left upon the sand, and distorted by the mist into a goblin. (*Lot* 200)

Then, abruptly, "the wildness of the feeling" passes and the fog begins to lift. Just as suddenly as these shapes emerged from the gloom, they are gone, and Curtis regains some command of the narrative. Now "the sea's wail, that wandered through the enchanted silence as if seeking sympathy" seems to resonate with his own mood, and he recalls Charles Kingsley's 1850 novel, *Alton Locke*, which offers some foothold in this scene that has just transpired. Wandering home, the night becomes "more merciful," the sound of a guitar floating through the air "like ghosts of sound haunting the moonlight" (*Lot* 202). With his vision restored, Curtis can once again make "pleasant" this evening on the beach, but only insofar as he can turn toward the stability of the literary archive and the conventions of a vacation. The fantastic clouds of the beach, the comings and goings of fog, come to figure as an interesting mystery only when they remain a minor moment within an otherwise pleasant seaside reverie.

Even with this shift back toward the rhetoric of a seaside vacation, Curtis cannot close the book with this strange, unsettling evening. Compared to his pleasant memories of Europe, these moods along the western Atlantic leave him unsettled, unsure of what might emerge from this rising fog of modernity. What monsters are these, appearing out of the Atlantic to unsettle one's sense of the beach, of the beautiful resort towns? Curtis does not say. Instead, he describes an experience that seems out of time with this night on the beach, a trip by boat out to an old fort, covered in "Turneresque" vines, where girls toss pebbles into the smooth waters and light plays on the ruins (*Lot* 204–5). Here, the scene once again serene and pleasant, he finds himself capable of closing the summer's tour, but only after first disconnecting the narrative present from the disruption of the fog, from those darker mists that seemed, briefly, more than an inconvenience. Disconnecting and dismembering his present moment, Curtis is able to leave us with a shining visit of Newport, "the last picture of a pleasant summer," but ultimately, *Lotus-Eating* remains fixed in time and space, the "Howadji" stuck in his memories of a former sojourn in Europe and unable to command a firm grasp on the landscape before him, or of the oncoming fog (205). In his pursuit of an American escape, an interlude from all of the demands placed on a writer and editor at the beginning of his career, he can see only a less burdened, more carefree period in his life.

In his writing and in his work as an editor, Curtis, like Thoreau, pressed up against the confines of mainstream travel writing, defining and in the process limiting the genre's scope. As an author he gave voice to early tourism in the northeastern United States, contributing to an industry wherein periodical writing would connect urban centers to rural communities throughout New York, New England, and eventually the rest of the country and the world. As an editor he effectively silenced Thoreau's quite different sense of travel writing. Treading the uncertain ground of a littoral zone, however, Curtis's confident prose, his critical vocabulary and cultured outlook, give way to a profound uncertainty. By falling back on the discourse of popular travel writing, he eschews the unsettling nature of travel, the sense of returning home with an entirely new sense of oneself, of the world, that will mark the excursion form. After leaving the editorship at *Putnam's*, he would turn to "armchair" writing and to politics, never again writing as the "Howadji."

Thoreau, meanwhile, would be left grasping at the pieces of two now entirely unmarketable essays. These manuscripts, cut off by *Putnam's*, suddenly offered a chance to experiment. "Yankee in Canada" had suggested an alluring sense of the great American interior, but it had also suggested a strikingly different cultural backdrop than he'd described in his first book, *A Week on the Concord and Merrimack Rivers*. Ultimately, "Yankee in Canada" had, indeed, been something of a failure, at least in terms of continued success within major periodicals. Instead, though, it gestured toward a method, a connective process wherein experience, writing, and reading could come together to offer some entirely new understanding of the world right before us. In turning to *Cape Cod*, now ripe for revision, Thoreau would come to fully embrace the sense of disturbance that marks the mature excursion form. In that text, and later in *The Maine Woods*, his process would come to hinge upon not merely traveling and then writing, but also thinking seriously about what those journeys meant, and about revisiting and reseeing those places, those experiences. As he would write to Blake in the fall of 1857, the key in drafting an excursion is "returning to this essay again and again, until you are satisfied that all that was important in your experience is in it."[35] Only then, after a long, arduous process of writing, reflection, and re-vision, can an excursion reflect that sense of instability and disturbance Thoreau first saw at the Falls of St. Anne. Reading "Yankee in Canada" and *Lotus-Eating* together with an eye toward these changes, we witness two contrasting means of navigating a dynamic planet in literary writing, ways of living in a world in which ecology, affect, human history, and environmental form are not only dramatically entangled, but radically unstable and fundamentally disturbed. In these two texts, we witness nonfiction worlds—and emergent literary forms—veering toward something new. We also see the emergence of Thoreau's mature excursion form, which finally comes to fruition not in "Yankee in Canada," but in the narrative that he worked on simultaneously, *Cape Cod*.

3

Affect and Ecological Form in *Walden* and *Cape Cod*

A man must generally get away some hundreds or thousands of miles from home before he can be said to begin his travels. Why not begin his travels at home? Would he have to go far or look very closely to discover novelties?

—Thoreau, *Journal*, August 6, 1851

At the conclusion of *Cape Cod* (1862), Thoreau's prose veers suddenly, ironically, toward the rhetoric of a travel guide. After suggesting a means of getting to the outer region of the Cape and a place to stay, the narrator offers some advice on when to travel. "Most persons visit the seaside in warm weather," he writes, "when the fogs are frequent, and the atmosphere is wont to be thick, and the charm of the sea is to some extent lost" (*CC* 214). He goes on to suggest what comes to seem like a very different kind of trip, in which the atmosphere and the moods of the ocean appear to be the primary attraction:

> But I *suspect* that the fall is the best season, for then the atmosphere is more transparent, and it is a greater pleasure to look out over the sea. The clear and bracing air, and the storms of autumn and winter even, are necessary in order that we may get the impression which the sea is calculated to make. (214)

At first glance, these suggestions might seem like a calculated jab at contemporary travel narratives and their relationship to fashionable resort communities, or perhaps just a dark joke—*Cape Cod* is hardly that kind of travel book, and no reader, having made it this far, would turn to these pages for recommendations. The book opens in the aftermath of a horrific shipwreck, a cast of "wreckers" untangling bloated human bodies from the valuable seaweed washed up on the beach, mourners and curious locals looking on. The narrator, at first seemingly unaffected by this scene, will come to dwell on death, frailty, and social decline for the remainder of his trip. Indeed, he never does describe those "sunny lives on the sand" that, he supposes, "most men" love to imagine (*CC* 212). Still, there is a sense of earnestness here at the book's conclusion that is difficult to shake—what *is* "that impression which the sea is calculated to make?" And what kind of a trip is Thoreau describing, which begins amidst such tragedy? What, in the end, are we meant to take away from these travels, if not a few suggestions?

Cape Cod was partially published in *Putnam's Monthly* in 1855, where unsuspecting readers encountered it as an anonymous literary travel essay. When the editor George William Curtis abruptly cut the piece after publishing the first four chapters, just as he had two years before with "Yankee in Canada," Thoreau continued to work on the manuscript, adding additional experiences from subsequent trips and extensive research in what he called the "ante-Pilgrim" history of New England (179). Thus, while the narrative perhaps began as a straightforward account of a single trip in 1849, the book, published more than fifteen years after that first walk on the beach, came to be something quite different. In his process of composition and revision, Thoreau suggests a literary method through which travel offers a new, unsettling sense of a world in motion, threatening to disrupt other, more human-centric stories—like that pleasant summer vacation that Curtis resorts to in *Lotus-Eating*. Revision and strategic addition, drawing from additional experience as well as significant outside research, came to be essential to Thoreau's mature excursion form. Of the four books that he published or nearly finished during his lifetime, *Cape Cod* is among the best examples of disturbance and adaptation, especially when examined alongside *Walden*, which he drafted and revised simultaneously.

This chapter turns to *Cape Cod* and *Walden*, focusing on what I call "environmental form" and the dual affects associated with disturbance and adaptation—the optimistic new dawn at the conclusion of *Walden* and the darker, more unstable sense of nature, economies, and US history which emerges in *Cape Cod*. The first section focuses on the composition and revision of *Walden* in the years following Thoreau's famous experiment in simple living, during which his regular walks and river trips came to function as short, near-to-home excursions. In one of the most famous chapters of *Walden*, "Spring," the narrator encounters a thawing sandbank, which ultimately suggests a new form, even a new language, more fit for the ecological present. This moment, which emerges from the practice of excursions, suggests both the utility of these daily trips beyond the home for a working writer and, at the same time, the startling discoveries that are central to the excursion form. *Walden*, I ultimately argue, thus draws heavily from the excursion form even while it is not entirely a travel narrative. The second section examines *Cape Cod* and the fundamental instability of both natural forms, such as beaches and dunes, and human culture, outlining three key ways in which this text anticipates Anthropocene nonfiction. Ultimately, *Cape Cod* offers both a template for embracing the most disturbing aspects of an excursion *and* the most direct departure from historical travel writing of any of Thoreau's books. Indeed, *Cape Cod* is exemplary of the mature excursion form because in moving out and away from a more settled way of seeing the world, the text dwells on earthly dynamism and ecological form, national myths and previous modes of writing and navigating the world fading away like Thoreau's own tracks on the beach. Reading these texts together suggests both the most unsettling effects of modern travel and the central importance of home in grounding the mature excursion form.

Walden and the Language of Wildness

When Thoreau set out for Canada in the fall of 1850, he was at the beginning of what scholars have widely regarded as a crucial turning point in his literary career, a moment when his prose became firmly grounded in observations of the material world. In September 1850, his Journal records a list of places within a day's walk of his home in Concord, familiar haunts including Walden Pond, Baker Farm, the Great Meadow, and more (*PJ* 3:23–4). In the months that follow, the Journal blossoms into a series of detailed daily outings on foot and by boat, and he would reflect regularly on both specific techniques for traveling, as well as the relative merits of remaining close to home, that great "paradox," first identified by John Aldrich Christie, between wanting to travel widely and staying home, where his observations reflect a detailed, intimate knowledge of the landscape.[1] *Walden*, too, reflects this shift. Not coincidentally, these excursions near home also coincide with other travels much further afield, including his time in Canada and trips to both Cape Cod and Maine, as well as many other excursions near and far which, while recorded in his Journal, were not published as standalone narratives. "Transitions are hard to mark," writes Laura Dassow Walls, "but one can take this list as the moment when Thoreau began to reinvent himself as the writer of world fame."[2] From this moment on, daily "travels" would become central to his method as a writer and thinker, and travel writing and natural history his primary genres. While trips beyond home feature prominently in many of his earlier works, including *A Week* and "Walk to Wachusett," as well as "Ktaadn," his mature excursion form is grounded in learning to travel, as he puts it in *Walden*, "a good deal in Concord" (*Wa* 4).

Scholars have recognized this shift for generations now, characterizing it as both a major stage in the composition of *Walden* and, more recently, as part of a broader "turn to science." Thoreau conducted his experiment in simple living from 1845 to 1847, during which time he worked on both *A Week* and *Walden*. According to Linck Johnson, he completed a draft of *A Week* "sometime in the fall of 1845," and though the book continued to, in Thoreau's words, "swell" with new material, Emerson urged him to publish as soon as possible.[3] By 1849, when *A Week* appeared in print, Thoreau was nearly finished with a draft of *Walden*, but he delayed publishing his second book, largely due to the commercial failure of *A Week*. According to foundational research by J. Lyndon Shanley, Thoreau completed seven drafts of *Walden*, composed in two main periods from 1846–9 and 1852–4.[4] William Rossi has recently complicated this narrative by examining Thoreau's Journal alongside his additions to the text, finding significant evidence of "persistent labor over the long haul," including more than sixty paragraphs drafted during "the supposed two-year hiatus."[5] Still, this moment does reflect both a renewed attention to the manuscript and a notable shift in Thoreau's prose style, and the later additions are grounded in a much greater density of environmental observation.

This shift in Thoreau's prose style also overlaps with and reflects an increased engagement with professional natural history, which scholars have often characterized as a "turn to science" or a "conversion to science." Influenced by his reading of Alexander von Humboldt, Charles Darwin, Asa Gray, and Jacob Bigelow, among others, Thoreau would ultimately shift away from his earlier idealism and resist symbolic views of nature, developing in the process a much more recognizably modern (and ultimately ecological) literary voice.[6] After 1850, his writing reflects an increasing engagement with science. James Perrin Warren, for example, traces Thoreau's interest in botany through specific scientific practices, including "observation, notation, reflection, writing, reading, and study in scientific texts."[7] During his regular excursions in and around Concord, as well as his trips further afield, Thoreau practiced detailed and systematic observation, methodically recording his findings in field books and later in the Journal, where they appear as vivid anecdotes, often with a broad cast of nonhuman characters and an uncommon awareness of seasonal and longer-term changes. *Walden*, the product of this transitional moment, ultimately reflects both the ambitious idealism of Thoreau's early writing and, through these empirical observations, a blossoming attention to what environmental philosopher David Abram calls the more-than-human world.[8] As Lawrence Buell notes, the composition of *Walden* demonstrates Thoreau's development as an environmental writer, a "path from homocentrism toward biocentrism," which finally culminates in "an environmentalist commitment."[9]

Walden is not an excursion in the same sense as *Cape Cod* and *The Maine Woods*, and it is certainly not a travel essay in the vein of "Yankee in Canada." Nonetheless, several scholars have observed that *Walden* is in its own way a unique kind of travel book. Robert Sattelmeyer argues that *Walden*, "in many ways the antithesis of a travel book and a celebration of staying home," is also "quite literally as well as hyperbolically about one who 'travel[s] a good deal in Concord.' "[10] John Aldrich Christie, meanwhile, is more decisive on this point, writing that "no book that Thoreau wrote made more effective structural use of the analogy of travel" for, as he puts it, "to write a travel book about a traveler who explores one place thoroughly, that he may thereby discover the resources of his own nature and spirit, is only the logical extension of the metaphor toward which all Thoreau's travel experience pointed."[11] Lawrence Buell also notes that Thoreau's book-length works are each built around one or more excursions, including, in *Walden*, "two years bivouacking at the pond."[12] Nonetheless, it remains useful to separate *Walden* from Thoreau's other, more far-ranging travel writing, as the text draws from a lifelong familiarity with its environment even as it finds new ways of seeing the landscape. Moreover, the narrator of *Walden* shifts the physical location of "home" throughout the text, from his familial home in the village to the retreat at the pond and finally back to the village. Of all his works, *Walden* is the most striking example of how Thoreau, as Walls puts it, "extends" the genre of travel writing "to compel an *inner* exploration in tandem with the outer."[13]

While his other books involve much lengthier travels, *Walden* relies on the excursion form in two significant ways: first, in its narrative and philosophical framing, and second, through the specific practices by which Thoreau gathered his material. In the opening lines of the book, the narrator presents a complicated sense of "home" which contrasts with the specific departures of his other travel writing. Here, he describes living in a "house" he built himself, where he worked at his writing. "At present," he adds, "I am a sojourner in civilized life again" (*Wa* 3). The figure of a "sojourner," a wanderer, suggests further movement even as the notion of "civilized life"—as contrasted with life in the woods—underscores a playful interrogation of what it means to live somewhere. Indeed, the physical site of home shifts from the village to the house at the pond and finally back to "civilized life again." Moreover, the narrator folds this structure of departure and return in on itself, for he also establishes his house as a site to begin one's travels, a place from which the meaning of journeys can take their final form. Thus, the larger departure and return to "civilized life," or simply life in the village, become the site of many other, smaller journeys. Indeed, the village itself becomes a destination.

By establishing his house at Walden as a site from which to set off on daily excursions, Thoreau also sets up his "business" there in terms of travel. "I have traveled a good deal in Concord," he quips in the opening pages (*Wa* 4). The account of his experiment which follows is painstakingly detailed, but the purpose is intentionally vague, even mystical: "to improve the nick of time;" to recover "a hound, a bay horse, and a turtle dove," all lost but never *quite* beyond reach; and ultimately to "anticipate" "Nature herself" (*Wa* 17). The only literal practices which connect these pursuits are the act of travel "in any weather, at any hour of the day or night" and, afterward, of writing, which he alludes to in terms of carrying the wind "express" and writing in a journal "of no very wide circulation" (*Wa* 17, 18). Moreover, the only other people who figure into this quest, besides the narrator himself, are themselves "travelers" with whom he has spoken (*Wa* 17). This playful reframing draws not merely from Thoreau's immediate context—nineteenth-century travel narratives and scientific expeditions—but also from antiquity, and especially from the pilgrimage, to suggest the central importance of travel as a *method* and a process of meaning-making. In not merely rejecting or pushing back against the bounds of contemporary travel literature, as he will in *Cape Cod*, but drawing from a longer tradition of travel as a site of meaning-making, Thoreau casts his experiment as a very modern quest, a search not for a specific place or a new home or some other material treasure, but rather for a new way of navigating a rapidly changing world.

Thoreau's playful experimentation, drawing from the heroic quest but peppered with irony and interrogation, belies the pragmatic nature of his experiment and the grounded nature of his craft. As *Walden* progresses, specific excursions take on increasing significance. While the early chapters focus on building the house and establishing its position as home, the later chapters draw heavily from anecdotes gathered while away. The opening paragraph of "Solitude," for instance, describes a walk along the shores of the pond on a "delicious evening" (*Wa* 129). Later, the narrator notes that "every day or two" he would walk to the village, now a destination rather than a point of departure (*Wa* 167). The chapter titled "The Ponds" presents Walden as well as other local ponds as regular destinations, the narrator noting that sometimes he "rambled still farther westward than I habitually dwell, into yet more unfrequented parts of the town" (*Wa* 173). The exchange with John Fields in "Baker Farm" features a fishing excursion and a visit with a neighbor, and the next chapter, "Higher Laws," begins with a return to the house at the pond and then a reflection on the purpose of setting out to fish or to hunt in the first place. Following this, episodes from daily walks feature prominently in each chapter. "Brute Neighbors" begins with an exchange with "the Poet," which sets up a familiar tension between staying home and engaging in intellectual pursuits and, on the other hand, setting out for another excursion (*Wa* 223). This chapter also features an episode with a loon, in which the narrator pursues the bird in vain, having first watched it from the home window and thus established a connection between experiences near the hearth and those just beyond the window. "House-Warming" also begins with a trip away, this time to collect wild grapes, with allusions to other foraging excursions for wild apples, chestnuts, groundnuts, and more. While it is an account of building a house in the woods, *Walden* is also filled with interesting experiences drawn from daily excursions. Indeed, these travels make up a far more significant portion of the text than his accounts of the house itself.

"Former Inhabitants, and Winter Visitors" provides several significant examples of these regular trips away. Excursions in this chapter reveal a history of Walden Woods as the home of an entire community of marginalized peoples, such as formerly enslaved African American Brister Freeman.[14] This history undercuts any notion of Walden Pond as an unpeopled wilderness, but it also establishes these woods as a site of refuge for diverse peoples who are left with nowhere else to go. Interestingly, this chapter combines these histories with accounts of winter walks, many of which seem to pivot around visits with other, nonhuman neighbors. For example, the narrator "frequently tramped eight or ten miles through the deepest snow to keep an appointment with a yellow-birch, or an old acquaintance among the pines" (*Wa* 265). He also recounts "watching a barred owl (*Strix nebulosa*) sitting on one of the lower dead limbs of a white pine, close to the trunk, in broad daylight, I standing within a rod of him" (*Wa* 266). When the bird finally takes flight, the narrator describes a new perspective on these woods, in which he imagines the owl being "guided amid the pine boughs rather by a delicate sense of their neighborhood than by sight, feeling his twilight way as it were" (*Wa* 266). Taken together, these encounters with previous human inhabitants and with nonhuman beings suggest the value of excursions in shaking the reader and the narrator alike to some new register. Faced with these unexpected new perspectives on a familiar environment, the narrative adapts. Indeed, the narrator's focus and voice have already begun to shift, the woods becoming not so much a site of retreat and secretive business as a place of "neighborhood" with humans, birds, and trees alike.

The penultimate chapter of *Walden*, "Spring," also draws from regular excursions and reflects these patterns of disturbance and adaptation. This chapter, and particularly the sandbank passage, is among the most celebrated moments in the entire book, called the "central idea" by Richardson and often considered the narrative climax.[15] First, the narrator sets up an image of a traveler in the springtime, navigating the landscape in the midst of the thaw:

> At length the sun's rays have attained the right angle, and warm winds blow up mist and rain and melt the snow banks, and the sun dispersing the mist smiles on a checkered landscape of russet and white smoking with incense, through which the traveler picks his way from islet to islet, cheered by the music of a thousand tinkling rills and rivulets whose veins are filled with the blood of winter which they are bearing off. (*Wa* 304)

This moment not only invokes the now familiar image of the narrator as a traveler in his home landscape, just beyond the bounds of normal experience, but also casts this spring morning in a mythical landscape, perhaps even outside the bounds of chronological time in a more natural, cyclical pattern demarcated by seasonal changes. This passage also marks the beginning of an extended metaphor in which the surface of the earth becomes itself a living body.

The traveler-narrator goes on to describe a process which is rooted in industrialized travel and anthropogenic changes on the land, but which ultimately reflects an organic vision of the earth and its myriad forms. "Few phenomena gave me more delight than to observe the forms which thawing sand and clay assume in flowing down the sides of the deep cut on the railroad," he writes. This was perhaps once uncommon, or difficult to observe at any large scale, "though the number of freshly exposed banks of the right material must have greatly multiplied since railroads were invented" (*Wa* 304). He describes the scene at length, sand and clay and water flowing together down the bank, a mix of colors and materials that is at first "like lava," but then "a sort of hybrid product, which obeys half the law of currents, and half way that of vegetation" (*Wa* 305).[16] The patterns which emerge suggest leaves, vines, and strands, a strange, primordial precursor to the foliage that might later emerge from this same material, as though the physical earth were here anticipating other, later biological forms. James Perrin Warren describes this profusion in terms of Thoreau's "plant thinking," or, in this case, the ways in which his writing and his thinking are shaped by a close attention to botany and to the particular lives of individual species and plants.[17] In this moment, however, Thoreau's interest shifts from plants themselves to material and form. If earth is literally the material here, and foliage the product and form of sand and clay and the other invisible materials therein, the narrator comes to see his own relation to the processes of creation and renewal that are at work in the deep cut.

During the right season, the narrator observes, this "sand foliage" is visible along great stretches of the railroad. But "what makes this sand foliage remarkable is its springing into existence thus suddenly" (*Wa* 306). In the lines that follow, he completes a movement from the banal through the biological and the geological and ultimately to the divine act of creation:

> When I see on the one side the inert bank—for the sun acts on one side first—and on the other this luxuriant foliage, the creation of an hour—I am affected as if in a peculiar sense I stood in the laboratory of the Artist who made the world and me—had come to where he was still at work, sporting on this bank, and with excess of energy strewing his fresh designs about. (*Wa* 306)

From here the passage becomes still more remarkable—for suddenly the narrator casts the planet earth as a great animal, and this sand foliage as a curious means of glimpsing into the inner workings of the earth as a living body, writing that he feels as though peering into "the vitals of the globe, for this sandy overflow is something such a foliaceous mass as the vitals of the animal body" (*Wa* 306). The leaf, however, remains a key, a primary form which ultimately connects the sand and vegetation with such wonders as bird wings, trees, rivers, and human language:

> No wonder that the earth expresses itself outwardly in leaves, it so labors with the idea inwardly. The atoms have learned this law, and are pregnant by it. The overhanging leaf sees here its prototype. *Internally*, whether in the globe or animal body, it is a moist thick *lobe*, a word especially applicable to the liver and lungs and the *leaves* of fat, (*γείβω, labor, lapsus*, to flow or slip downward, a lapsing; γοβος, *globus*, lobe, globe; also lap, flap, and many other words,) *externally* a dry thin *leaf*, even as the *f* and *v* are a pressed and dried *b*. The radicals of lobe are *lb*, the soft mass of the *b* (single lobed, or B, double lobed,)

> with the liquid *l* behind it pressing it forward. In globe, *glb*, the guttural *g* adds to the meaning the capacity of the throat. The feathers and wings of birds are still drier and thinner leaves. Thus, also, you pass from the lumpish grub in the earth to the airy and fluttering butterfly. The very globe continually transcends and translates itself, and becomes winged in its orbit. Even ice begins with delicate crystal leaves, as if it had flowed into moulds which the fronds of water plants have impressed on the watery mirror. The whole tree itself is but one leaf, and rivers are still vaster leaves whose pulp is intervening earth, and towns and cities are the ova of insects in their axils. (*Wa* 306–7)

This passage concludes with a remarkable revelation, which is ultimately rooted in the process of composing literary text from observations of natural phenomena. "Thus it seemed that this one hillside," the narrator concludes, "illustrated the principle of all the operations of Nature" (*Wa* 308). This insight, however, is not simply geological. As Walls writes, "As the raw wound of the railroad's 'deep cut' thaws and flows, he sees in the flowing sands the canvas of creation, revealing the great truth that we live not on the surface of a dead planet but in and through a living earth, like a leaf unfolding."[18] As the narrator puts it, "The earth is not a mere fragment of dead history, stratum upon stratum like the leaves of a book, to be studied by geologists and antiquaries chiefly, but living poetry like the leaves of a tree, which precede flowers and fruit—not a fossil earth, but a living earth" (*Wa* 309). The very text of *Walden*, he suggests, is part of the same process of creation, drawing its energy from the vital dynamism of the material planet—and from the narrator's many excursions.

Literary scholars have long considered this passage and its key line—"nothing is inorganic"—the "central idea" of *Walden* and a mark of Thoreau's "maturity" as a writer.[19] Research in the field of ecocriticism and in Thoreau's manuscripts has complicated this, however, suggesting instead an ongoing, somewhat irregular development of Thoreau's environmental thinking and writing. While *Walden* does represent a shift in Thoreau's style and focus, it was neither sudden nor particularly neat. Over the course of his life and in his specific revisions and additions to the manuscript, he gradually developed a method of centering nonhuman beings and ecological processes. Buell calls this process "an irregular movement" in which the "substantial reality" of the nonhuman world comes to supersede its aesthetic or didactic "use." *Walden*, he adds, does not contain Thoreau's most "self-consciously environmentalist statements, nor his most close-grained observations" of the nonhuman world, and even the book's most detailed descriptions of natural phenomena, including the sandbank, are "allegorized more aggressively" than his later writing.[20] In "Spring," for instance, the new season comes to represent a change in the narrator himself, a move which serves to at least partially obfuscate the transformations of the physical world even as it also places the narrator and his use of language within a much broader, more-than-human network of creative energy.

Moreover, this specific passage is not purely a late addition to the manuscript, and cannot be taken as evidence of a single, decisive pivot. Thoreau first drafted what Rossi calls the "core" of the sandbank passage in 1848, which, he notes, already "begins to assume the final form in diction, syntax, and structure."[21] Thoreau's revisions, Rossi explains, are characteristically "accretive," made by inserting new material into drafts that were already written out. He made a few minor, descriptive additions to the passage in the following two years, when, in the spring, the same phenomenon was once again visible on the landscape. These minor revisions, he concludes, were completed no later than December 1851. The majority of the passage's "lyrical generalizations," meanwhile, including "the elaborate analogies of 'leaf' and 'lobe forms,' rapid shifts of perspective from the granular to the planetary, and the final rhapsodic peroration to vital force and plasticity," are drawn from passages in the Journal after December 1851.[22] Thus the earliest, most materially grounded portions of this iconic passage actually predate Thoreau's supposed "turn" to science and the material world, while the later revisions draw from literary abstractions to enhance the meaning of the passage for human readers.

In the end, this moment in *Walden* seems to figure a new style of writing in order to suggest not a sudden shift in literary method, but rather a moment when the narrator is, as Walls writes, "remade, and ready."[23] His reaching for a new language in the sandbank passage is not an immediate pivot, but rather a reflection of the narrator's emergent commitment to such writing, to drawing from the material world in such a way as to suggest an entirely new sense of humans and our relationship to the world. The methods that encompass Thoreau's excursion form are emerging in real time as the narrator examines these natural forms and the forces, both human and inhuman, which shape them. For even as he looks inward, Thoreau also registers the remarkable potential for the material world to create meaningful experience, even in a mundane place, and despite being significantly reshaped by the railroad. We ought to take him literally, then, when he says that he has "traveled a good deal in Concord," for *Walden* does indeed draw from the genre of travel literature and from the specific practices of the excursion form. And while the book is perhaps as much memoir as travel narrative, and by turns poem, novel, sermon, and practical treatise, it does represent a significant shift toward the more unsettled characteristics of Thoreau's two later travel books, *Cape Cod* and *The Maine Woods.*

In a material sense, the sandbank passage prefigures Thoreau's growing interest in natural forms, the transformation of glacially deposited sand into a steep bank and then into something else. In observing and then depicting this formal shift, the narrator's playful language becomes like the sand, the *material* which, in response to creative energies, shifts from one form to another. Just as a wooded hill becomes a sandbank which gradually slumps and spreads over the edge of the tracks to form a miniature alluvial plain, so Thoreau's writing "continually transcends and translates itself" (*Wa* 307). *Walden*, like *A Week* before and much as the travel books which were even then beginning to emerge, is thus an experiment in form. As such, it suggests two key insights for Thoreau's method. First, regular, mostly local excursions became the bedrock of his writing, offering a framework and also the key insights of all his later works. Second and just as important, while these local excursions are often transformed almost immediately into literary vignettes in the Journal, the final meaning of these moments often emerged only later through extended revision and reflection. From this moment on, Thoreau's methods will be grounded in careful, deliberate mediation of experience, just as he would suggest to Blake in 1857. Thus, by grounding his work in direct observations of the material world, Thoreau was not only gathering material but creating a physical site for developing and refining the meaning of lived experience. *Walden* is thus key to the excursion form, an experiment which suggested a new means of bringing together literary writing and lived experience with a new, imminent sense of earthly dynamism. This key methodological insight is what allows him to develop a theory of literature which is grounded in disturbance and adaptation, but it was not, in the end, his most dramatic experiment in form. While *Walden* ultimately figures this vital energy as a rebirth, concluding with an optimistic image of dawn, *Cape Cod*, observing that same energy less than one hundred miles away, presents a strikingly different affect.

Affect, Ecology, and the Form of *Cape Cod*

A little over a year after Thoreau first described the phenomena he called "sand foliage" in his Journal, he traveled to the outermost reaches of Cape Cod.

There, walking the beach amidst the aftermath of a powerful autumn storm, he observed other natural forms in a state of constant motion. His descriptions of dunes, beaches, and turbulent weather reflect a remarkably different aspect of the creative energy he describes in *Walden*. *Cape Cod* comes to reflect a unique set of connections among atmosphere, beach ecology, and narrative form. While *Walden* suggests a new commitment to an emergent method, *Cape Cod* goes further, outlining three key features of the excursion form that are grounded in the emergent instability of the Anthropocene: first, the difficulty of seeing the modern world as it really is due to large, complex systems; second, the inhuman nature of climate and natural systems; and finally, the startling destruction and (re)creation of environments that humans rely on. Ultimately, the text critiques economic systems and nationalist histories of North America, suggesting that modern ways of living and being in the world are not only inhumane, but fundamentally out of touch with the actual conditions shaping the world, from the contingencies of history to the storms of the present. For readers today, *Cape Cod* offers a vital starting point for the new nonfiction literatures of the Anthropocene, at least as significant as *Walden*, because it draws from the dynamism of its material present to reimagine the form and style of both its environmental representations and critique.

Cape Cod has produced a wide variety of reactions among readers, and until recently, it has remained understudied, probably because it is so different from Thoreau's other works. Critics and biographers have variously called it a sunny, pleasant work of local color and, conversely, a macabre essay on human frailty in the face of nonhuman nature.[24] Katie Simon suggests that compared to *Walden* and *The Maine Woods*, the book's content and strange affect present an "obvious challenge" for "readers who perceive Thoreau as a heroic environmentalist."[25] Indeed, while *Walden* presents a pastoral retreat and, for many readers, a paean to the redemptive powers of nature, *Cape Cod* is notably different, by turns frank, pragmatic, brooding, and often dark but just as often sly and humorous. For Simon, this strange affect represents a "hauntological sublime," and the book "aims to disturb readers" in its "radical critique of broader economic systems underlying relations between nature and the human."[26] Whereas in *Walden* the narrator claims that he writes "to wake my neighbors up," *Cape Cod* is calculated to unsettle, challenge, and trouble (*Wa* 84). This disturbing nature of the text is a hallmark of Thoreau's mature excursion form—and one that entails a substantial risk, not only of misinterpretation, but of actively turning away readers.

Recent scholarship on *Cape Cod* has both explored the darker themes of the text and established its sophisticated construction, pushing back against earlier assumptions that the book was unfinished or uninspired. Robert Pinsky, for instance, calls *Cape Cod* a "sardonic, maniacally various work," at once "quirky, scholarly, thorny, clowning" and "full of startling jokes" because it was written to be performed at public lectures.[27] The wreck of the *St. John* has been a central point of inquiry for many scholars, as well as the critique of capitalism and its "remaking" of the human. For example, Alex Moskowitz argues that the juxtaposition of the wreck of the *St. John* and everyday economic concerns like gathering and selling seaweed creates a unique form of "economic imperception," in which the most affecting observations of modern life are registered only to be ignored.[28] Likewise, Jamie L. Jones reads *Cape Cod* as a resource for learning to see global capitalism and climate change, a text which "prescribes useful methods for trying to see injustice and power at sea—even though our attempts to understand everything at once will always be limited."[29] Another vein of scholarship has explored *Cape Cod* in terms of transnationalism and Thoreau's unique capacity to, in the words of Paul Giles, "interweave close and distant, near and far" in "a style of juxtaposition through which different spatial and temporal coordinates are brought productively into contact with each other."[30] Christa Holm Vogelius, meanwhile, argues that *Cape Cod* offers an image of the beach as "the site where the deep history of the American soil and the ahistorical wilderness of the ocean meet violently," a move which "undoes" the Transcendentalist nation-mapping of, for example, "Walking."[31] Additional research has explored *Cape Cod* in terms of its engagement with cartography and history.[32] Together, this body of scholarship suggests that *Cape* Cod reflects the then emergent global instability of the Anthropocene.

When he began drafting his account of that 1849 trip, Thoreau likely imagined it as another piece for the popular literary periodicals, much like "Ktaadn" and "Yankee in Canada." Ultimately, however, *Cape Cod* is so unsettling only because it reflects a significantly extended process of revision and reflection. Thoreau's narrator suggests in the opening lines that his narrative is drawn from three trips: "I made a visit to Cape Cod in October, 1849," he writes, "another the succeeding June, and another to Truro in July 1855; the first and last time with a single companion, the second time alone" (*CC* 3). Even this, however, belies the complexity of the final text. In fact, he made several trips to Cape Cod and the immediate vicinity, and he spent over a decade researching the history of the New England coast, incorporating new material gleaned from his readings until well after 1855. Indeed, while Thoreau famously copied historical maps to accompany his travels on the Cape, readers interested in literary craft might benefit from a map of the text and its many allusions, additions, and references to other documents. This is, to be certain, no simple account of a walk on the beach. Rather, *Cape Cod* compresses several trips into a single narrative, drawing from a variety of experiences both on the Cape and elsewhere on the coast.

The final version of *Cape Cod* that readers encounter today reflects a significant and complex system of revision spanning several years. The structure of the text is based on the 1849 trip, when he and the poet William Ellery Channing left Concord for Boston, planning to take the steamer to Provincetown and walk back toward the mainland. Instead, they encountered a gale which halted their travel and caused, near Cohasset, the wreck of a brig carrying Irish immigrants. Taking the train south from Boston, they observed the wreck before setting out on a walk that would eventually take them to the home of a retired oysterman, John Newcomb, past the Highland Light, and all the way to Provincetown at the tip of the Cape. They returned to Boston via the steamer *Naushon* six days later. This route and its stormy weather form the frame of *Cape Cod*, from the opening chapter in Cohasset to the dreamy return via the steamer, when Thoreau and Channing watch the Cape disappear behind them.

The two additional excursions Thoreau mentions, as well as another trip to a different beach in 1850, played smaller but nonetheless important roles in his writing process. The second trip, undertaken less than a year after the first excursion in June 1850, contributed significantly to the evolving essay. Traveling alone, Thoreau took the steamer from Boston to Provincetown, returning by the same means after five days exploring near Highland Light and once again visiting the oysterman, John Newcomb. Throughout the text, Thoreau refers to moments "the next summer" and "in June," introducing passages taken from his 1850 Journal account, when the weather was far more hospitable. He also began incorporating material from his reading at about this time, including many scientific and historical texts.[33] Although he doesn't mention it explicitly in *Cape Cod*, Thoreau also made an additional trip to another beach just a few weeks later, in July 1850, when Margaret Fuller, her husband the Marquis d'Ossoli, their young son Angelo, and several others drowned after the wreck of the *Elizabeth* off of Fire Island, New York. He had been sent by Emerson to recover the bodies of Fuller and her family, and if possible, her manuscript on the Italian revolution. Failing these, he left with only a button picked up from the sand and a record of the experience in his Journal. As Joseph Moldenhauer notes, this somber journey "contributed various reflections on death and shipwreck and a couple of anecdotes about Long Island watermen to the Cape Cod narrative."[34] Thoreau also made a trip to Plymouth and Clark's Island, near the Cape, in July 1851, which would provide a brief account of one "Uncle Bill," who had lived aboard a schooner on the shores of Plymouth Bay.[35] Thus by 1851, the Cape Cod narrative was beginning to encompass significant additional material, including substantial research as well as brief accounts and allusions to other experiences. While "Yankee in Canada" had been a "straightforward" account of a single, rather rushed excursion, within two years, *Cape Cod* was beginning to reflect a longer, more nuanced process of composition and revision, with significant additional experiences in the same locations and in the general area.

Throughout 1851 Thoreau worked on both the Cape Cod essay and his Canada excursion simultaneously, in addition to *Walden*. In November 1852, he sent the first three chapters of *Cape Cod*, "The Shipwreck," "Stage-Coach Views," and "The Plains of Nauset," to George William Curtis at the new *Putnam's Monthly*, "not yet half the whole," as he put it in the accompanying letter. The fourth chapter, "The Beach," was nearly ready as well (*Corr* 2:132). When the inaugural numbers of *Putnam's* appeared in 1853, however, Thoreau found his "Excursion to Canada," which had been submitted on his behalf by Horace Greeley and serialized in the January number. Thoreau was no doubt vexed to find two manuscripts caught up in the hands of a single editor, but he apparently reached an agreement with *Putnam's*. The situation grew worse, however, when Curtis cut those offensive lines from the Canada essay. Greeley, who had struggled to place the Canada essay and knew that any travel piece must be timely, attempted to persuade him to remain diplomatic. In any piece published anonymously, like all contributions to *Putnam's*, he urged, "the elimination of very flagrant heresies (like your defiant Pantheism) becomes a necessity" (*Corr* 2:139). Curtis, as editor, was merely mitigating risk. Thoreau, however, would not compromise. Communications broke down before the fourth installment, which would have gone to press in April. Still, Greeley urged caution. "Don't break with Curtis or Putnam," he wrote in a March 1853 letter (*Corr* 2:156).

In another letter from this same period, Thoreau hints to H. G. O. Blake that *Cape Cod*, caught in the midst of this disagreement, would be canceled as well. In fact, Thoreau took Greeley's advice after all, and *Cape Cod* remained in Curtis's hands for the next two years. The first four chapters were published serially in three installments in June, July, and August 1855. Interestingly, Thoreau continued to add new material even as these first installments were printed. That July, he made his third trip to the Cape, again with Channing, staying in a newly constructed hotel at the Highland Lighthouse for a week and taking short walks in the vicinity. Moldenhauer notes that the Journal passages recording this excursion, twelve manuscript pages in total, are predominantly "close observation of flora, fauna, and weather" as well as "information about fishing and erosion elicited from Cape residents." Most of these passages were almost immediately incorporated into the manuscript, even "as he fleshed them out in the journal notebook from rough field notes."[36] Working quickly and efficiently in what seemed a decidedly new style, Thoreau was drawing from the same methods that contributed to the later revisions of *Walden*, including much richer descriptions of his environment.[37] He expected *Cape Cod* to continue apace, anticipating that the next chapter, "The Wellfleet Oysterman," would appear in the September issue. He even wrote to Curtis in early August requesting a revision in the chapter which would follow, "The Beach Again," in which he substituted material gleaned from his 1855 trip.[38]

That new material, however, would not ultimately appear in the September issue. Despite the editorial conflicts, Thoreau had intended to publish the remainder of *Cape Cod* in *Putnam's*, but Curtis informed him late that summer that the series would be canceled. According to Moldenhauer, the decision was likely based on "the tone and content of 'The Wellfleet Oysterman,'" plus the previous conflicts over "Excursion to Canada."[39] Moreover, the series had already drawn negative reactions on the Cape. Curtis seems to have known that the next section of the manuscript, which included the chapter "The Wellfleet Oysterman," would cause even more controversy among readers, with its gritty, spitting portrait of one otherwise hospitable and deeply interesting Cape Codder. Like "Yankee in Canada," *Cape Cod* would be published in its entirety only after Thoreau's death. Still, he continued to work on the manuscript.

After *Putnam's* canceled the series in 1855, Thoreau made four additional trips to the seashore. In June 1856 and again in April 1857, he visited his friend Daniel Ricketson in New Bedford, each visit contributing a minor anecdote to *Cape Cod*.[40] In June 1857 he made an additional excursion to the Cape, the account of which fills some fifty pages of the Journal, narrating a walk from Manomet Point near Plymouth all the way to Provincetown, navigating by map and compass and often avoiding established paths and roads. This trip came at a significant moment for Thoreau, who had been suffering from a mysterious illness and unable to take his customary walks in 1855. Indeed, his excursion to the Highland Lighthouse in June of that year was partly an effort to recover his health and vigor. The 1857 excursion was marked by a return of his old energy, not to mention his habits of rambling across lots and away from main roads. It was also a precursor to his third trip to Maine later that year, which is the subject of "The Allegash and East Branch," the third essay in *The Maine Woods*. Interestingly, and despite the detailed accounts in his Journal, this final excursion to the Cape provided only very little new material for *Cape Cod*.[41] Although he continued working on the manuscript at least occasionally after 1857 and up to his final days, he focused much of his attention on other essays, only adding historical material from his readings to the final chapter, "Provincetown."[42]

Thoreau's composition of *Cape Cod* came to resemble the painstaking, drawn-out process that produced *Walden*, wherein experiences over an extended period of time are layered and compressed into a single final narrative, often with significant veers along the way. Thus, while the 1849 trip provided the basic narrative structure, the text itself reflects, to varying extents, several other excursions in and near the Cape, not to mention more than a decade of reading and research. To map Thoreau's writing of this excursion is thus to trace an elaborate process of accretion, beginning with a sustained narrative and drawing from a variety of additional anecdotes gathered from other points and times and an exhaustive list of references to other writers and historical documents. The end result is an excursion that emphasizes, rather than the destination or the routes, the subjective meaning of the traveler's lived experience. This method of composing travel literature, of reflecting, returning, and ultimately emphasizing certain aspects of experience, is key to Thoreau's mature excursion form and its utility for contemporary travelers. Moreover, the patterns of disturbance which shape this process, from shipwrecks to locked humane houses and the shifting ecologies of the beach itself, ultimately dictate the very shape and form of the text.

From the start, *Cape Cod* is shaped by weather, though the most haunting aspects of the text at first seem less meteorological than cultural and economic. The opening chapter, "The Shipwreck," famously begins amidst the wreckage of the *St. John*, when the narrator observes the collected bodies of more than two dozen Irish immigrants being nailed into coffins even as more victims were being washed ashore. As he puts it, there were "no signs of grief, but there was a sober dispatch of business which was affecting" (*CC* 5). Even more unsettling, professional "wreckers" scoured the beach for anything of value that had washed ashore, and seaweed collectors gathered a "valuable manure" in the wake of the storm, which they even untangled from dead bodies and scraps of clothing. The narrator casts himself as something of a wrecker in this passage, seeking, if not material goods, then the *material* for his narrative. Thus, he implicates himself in a critique of our relentless drive to accrue capital, for if the onlookers are unaffected by the tragedy before them, he too is shockingly unmoved, and the entire "fabric of society" will continue, in his words, without "a visible vibration" (*CC* 7).

But as Walls observes, there *are* reverberations, as when the narrator notes that "a man can attend but one funeral in the course of his life, can behold but one corpse" (*CC* 9). This is, as Walls notes, a cryptic allusion to Thoreau's brother, the great loss of his life, another "St. John."[43] Indeed, it seems as though affect cannot ultimately be repressed, at least not entirely, and not for long. If "Yankee in Canada" produces a sticky sense of something more just beyond the present horizon, *Cape Cod*—written simultaneously and drawing from and commenting on many of the same historical sources, the notes even intermingling across the texts—produces something which adheres still more tightly. Thus, while the reader is initially revolted by the narrator's cold observations of "many marbled feet and matted heads" and "one livid, swollen mangled body of a drowned girl," the broader social and economic system in which the wreckers operate is ultimately far more unsettling, shaping all aspects of the text (*CC* 5). Only as the narrator walks along the beach, observing the bleak conditions there, the depleted soil and lack of tree cover, do we come to see the overwhelming necessity of wrecking, of scratching out a living despite everything, surrounded by a cruel and unforgiving economic and social system. In the process, this travel essay becomes not so much the report of a trip as a map toward a broader social and economic awareness, a means of reaching toward a much broader affective and spatiotemporal register.

In the chapters which follow this horrific scene, *Cape Cod* draws from and attempts to make visible the emergent instability of the Anthropocene in three ways. First, the narrator reflects on the difficulty of observing the world. In Concord or some other, more frequent haunt, his layered observations over many seasons and years suggest the regular recurrence of specific phenomena, such as the sand foliage or the bloom of plants or nesting of birds. Here, however, lacking that deep knowledge, the narrator struggles to perceive both specific objects and the larger systems which influence them. Walking on the beach challenges his sense of distance, and objects on the sand seem to loom and then shift before his very eyes. At the Highland Light, the narrator calls the Cape "a place of wonders," where a nearby sandbank is transformed into a distant mountain range, luring mariners to run their vessels aground (*CC* 137). He describes yet more mirages near Provincetown, where a series of freshwater ponds seems to be set "by magic" on the side of a hill, and where, at sundown, a ridge of sand some twelve feet high appears to rise parallel to the shore (*CC* 150–2). These mirages also suggest a haunting. The narrator recalls a moment on Fire Island, when, searching for the body of Fuller, he was directed to human remains on a remote section of beach, just "some bones with a little flesh adhering to them" (*CC* 84). He expected them to be difficult to find, but in fact they loomed large as he approached. Then, coming nearer, they came to seem insignificant, and as he recalls "there was nothing at all remarkable about them, and they were singularly inoffensive to the sense and the imagination" (*CC* 84). But a moment later, standing there with the wind roaring and the surf crashing, the remains grow "more and more imposing" until the body seems to haunt the place, "as if there was an understanding between them and the ocean which necessarily left me out" (*CC* 84). For someone like Thoreau, obsessively recording precise observations, the Cape proves a bewildering environment, both impossible to observe and darkly suggestive. By returning to these difficulties throughout, the narrator thus highlights the importance of *trying* to observe the material world in its current state. The text thus anticipates later Anthropocene nonfiction, when larger, less directly observable phenomena come to loom constantly over all human art and literature, from colonialism to climate tipping points and feedback loops.

The second means by which *Cape Cod* anticipates later Anthropocene nonfiction is through its focus on weather, which ultimately comes to not only suggest a mood but also to physically shape both the material world and the text itself. As in twenty-first-century accounts of extreme weather and climate breakdown, Thoreau's prose evinces what seems an obsession with weather, even when the narrator's focus seems to be elsewhere. The storm which sank the *St. John* and forced Thoreau to change his plans is like another character in the text. Indeed, even when conditions are calm, the narrator muses that the ocean can almost instantly "be lashed into sudden fury, and all its caves and cliffs resound with tumult" (*CC* 98). In the final few chapters, nearer the end of the Cape, rain gives way to a fierce wind, and the narrator and his companion are "frequently glad to take shelter behind a sand-hill" (*CC* 145). Ultimately, though, this wind, a "New England nor'easter," is inescapable even behind the dunes, "for it blew about them, rounding them into cones, and was sure to find us out on whichever side we sat" (*CC* 160). These descriptions of biting wind and cold weather continue to grow through most of the narrative, setting a scene which, the narrator admits, cannot ultimately be captured by literary prose. "But what is our account?" he muses in the final pages. "In it there is no roar, no beach-birds, no tow-cloth" (*CC* 212). This question is significant because it suggests a reaching toward new modes of writing, a means of registering climate and its constant effects on human life and narrative. At the same time, by gesturing toward the insufficiency of literary narrative, Thoreau also subtly foregrounds the primacy of lived experience, in which meaning unfolds amidst a broader, more-than-human register. Walk the beach, in other words, and this obsession with weather, with the potential for a violent remaking of the ground beneath one's feet, seems like the only appropriate response.

The final way in which *Cape Cod* reflects an emergent Anthropocene style is through its insistence on the fundamental dynamism of the planet itself, particularly the beach. Indeed, the text *dwells* on the instability of littoral zones, those places where land and water mingle. This, too, begins in the opening chapter but seems to come to a head the further the narrator walks, until finally, in "Provincetown," he observes that "the sand is the great enemy here" (*CC* 173). If the beach is constantly being shaped and reformed all along the Cape, here the dunes are shifting year by year, even season by season, constantly threatening to inundate human structures and livelihoods. And while some places in Provincetown might be swallowed by the sand, others are in danger of being swept away. Indeed, the Highland Light, a major landmark in *Cape Cod*, was rebuilt in 1857 and then moved 450 feet away from the beach in 1996.[44] If the weather thus suggests a mood, the shipwreck at Cohasset reverberating in the dark, stormy weather and the biting, unshakeable wind, readers encounter "Provincetown" as literal quicksand, a mire of eroding beaches, encroaching dunes, and forgotten and alternative histories, which Thoreau incorporates into the text, forcing his reader to slog through the beach, crouching in the wind, for many more pages than any previous chapter. Michael Weisenberg identifies in this chapter an "alternative, cosmopolitan history" that ultimately reconsiders and complicates the cultural narratives inscribed upon the land.[45] Even the past cannot provide a stable footing on this landscape.

By its final chapter, Cape Cod has come to seem a dramatically unsettled landscape, a place in which the very ground beneath our feet shifts with every step, where the atmosphere threatens human life at every moment, and where the traveler cannot trust his own observations and instincts. Along the way, the text has steadily undermined our expectations. The sudden turn in the final pages toward the rhetoric of a travel guide, then, is all the more striking, for here, Thoreau both undermines the travel industry and comes to speak directly toward an emergent future, drawing from that same vital energy that he saw in the sandbank near Walden Pond. In the final paragraph of *Cape Cod*, the narrator looks toward a future when the Cape might be a "fashionable" destination:

> The time must come when this coast will be a place of resort for those New-Englanders who really wish to visit sea-side. At present it is wholly unknown to the fashionable world, and probably it will never be agreeable to them. If it is merely a ten-pin alley, or a circular railway, or an ocean of mint-julep, that the visitor is in search of—if he thinks more of the wine than the brine, as I suspect some do at Newport—I trust that for a long time he will be disappointed here. But this shore will never be more attractive than it is now. Such beaches as are fashionable are here made and unmade in a day, I may almost say, by the sea shifting its sands. Lynn and Nantasket! this bare and bended arm it is that makes the bay in which they lie so snugly. What are springs and waterfalls? Here is the spring of springs, the waterfall of waterfalls. A storm in the fall or winter is the time to visit it; a lighthouse or a fisherman's hut the true hotel. A man may stand there and put all America behind him. (*CC* 214–15)

This passage is, of course, a darkly ironic critique of the travel industry—and of the popular travel writing of the day. In his comments on Newport, Thoreau's narrator makes a sly jab at Curtis, whose *Lotus-Eating* describes a pleasant seaside vacation there. For Curtis, this "thick" atmosphere is a problem insofar as it disrupts an otherwise pleasant vacation, suggesting London smog or, interestingly, a great looming monster. For Thoreau, though, a vacation was never the point. By opening with the wreck of the *St. John*, he establishes the urgency of *Cape Cod*'s haunting, and the weather, the strange apparitions, the dynamism of the beach are not an "inconvenience" to be overcome by literary device, but a revelation of a stark and dangerous world and the human and inhuman forces shaping it. His advice for would-be travelers thus enacts a reversal in which human-centered narratives gradually fade away amidst larger ecological processes—beach formation and erosion, seasonal shifts in climate, botanical succession. By turning away from the vacation and toward other kinds of stories, toward that "impression" arising from the sea on clearer, colder, and stormier days, Thoreau confronts the same creative forces that he saw on Katahdin and in the thawing and freezing sandbanks near Walden Pond. For readers today, he suggests not only new kinds of narrative, but also new kinds of travel, new ways of creating meaningful experience on a dynamic, damaged planet. This is the Thoreau who speaks most directly to the twenty-first century—the experimental writer who aims to unsettle our easy ways of telling a story.

4

George Copway's Figures of Homesickness

I view my life like the mariner on the wide ocean, without a compass, in the dark night, as he watches the heaven for the north star, which his eye having discovered, he makes his way amidst surging seas, and tossed by angry billows into the very jaws of death, till he arrives safely anchored at port. I have been tossed with hope and fear in this life.

—George Copway

While Thoreau was developing his mature excursion form, other authors were also experimenting with travel writing as a means of navigating the dramatic changes shaping the American continent. Writing in his autobiography, *The Life, History, and Travels of Kah-ge-ga-gah-bowh* (1847), Mississauga Ojibwe author George Copway looks back at a much-anticipated homecoming.[1] For more than five years he's been traveling, teaching, and preaching among Ojibwe communities near Lake Superior and in the upper Mississippi River basin—and the whole time he's longed for home. "Never did I feel so rejoiced as when I stood on top of a hill, and saw my village, seven miles across the lake," he writes (*Life* 125).[2] He goes on to describe a complex surge of emotions:

> I gazed upon it in pure delight; and as I took a retrospective view of all the scenes which I had passed through, I wondered at myself, and at the great goodness of God . . . . While crossing the lake, I was in perfect ecstasies; my heart leaped with joy; and my thoughts and emotions were at my home long before my person. O how tedious and tardy the boat ride seemed to be; I wished for wings several times. But at last, I planted my foot upon the spot on which I had been reared from infancy, and where some of my sweetest and happiest recollections of my life were centered. But "every sweet has it [*sic*] bitter." (*Life* 125)

Copway's prose often draws from the conventions of sentimental Christian life writing, but this moment stands out for its poignancy, lending a particular emphasis to his ecstatic return to Rice Lake. Arriving again in Rice Lake, the reader senses, marks a threshold moment in the travels of Kah-ge-ga-gah-bowh, a turning point.

While Thoreau was developing an excursion form which highlights the creative energies of the dynamic, contested planet on which the traveler walks, George Copway utilized travel narrative as a means of navigating a rapidly changing sociopolitical world. The differences are crucial. Thoreau's excursions, as well as *Walden*, are predicated on a relatively stable sense of home, that "bedrock" to which the text and the traveler always return. Copway, too, deploys an out-and-back narrative structure, but ultimately finds himself radically unstuck, suddenly and dramatically capable of moving around in the world but also dissociated from his Indigenous connection to a specific geographic place and the community that remains there. *Life, History, and Travels*, while thus not precisely an excursion in the sense of Thoreau's *Cape Cod* and *The Maine Woods*, provides Copway a means of connecting these seemingly disparate phases of his life, but it also bears testament to the extreme danger facing Indigenous communities in the nineteenth-century United States and Canada.

Copway is a somewhat enigmatic figure in the field of North American literature, and despite achieving remarkable literary success in his day, his work remains understudied. Born in Canada in 1818, Copway traveled in the present-day states of Michigan, Wisconsin, and Minnesota, where he served as a teacher, translator, and missionary to Ojibwe peoples living near the present-day communities of Keweenaw Bay, Bad River, Red Cliff, Lac Courte Oreilles, and Fond du Lac. Beginning in 1847, he briefly achieved celebrity status in the United States before descending into debt and obscurity in the 1850s. In this chapter, I read *Life, History, and Travels* in terms of Copway's witness to a series of dramatic sociopolitical and environmental changes taking place in the nineteenth century, including the arrival of Protestant religious cultures in the Great Lakes region, the decline of the fur trade, the beginning of an American timber and mineral boom in Michigan and further west, and, most significantly, the emergence of the Ojibwe Nation as a collective political entity. Cathy Rex has called his autobiography a means of "experimenting with forms of survivance and cultural fluidity."[3] I agree and take this one step further, interpreting *Life, History, and Travels* specifically as an experimental travel essay marked by arrivals and sentimentally charged homecomings.

This chapter begins with an overview of Copway's historical moment, focusing on the notion of an Indian Nation, a concept which he used to describe Ojibwe communities and political councils in the United States and Canada. During Copway's lifetime, the Ojibwe, previously scattered and politically disjointed by the fur trade, emerged as a collective political entity, though not without friction. For example, Copway may have been perceived as something of an outsider among Lake Superior and Upper Mississippi Ojibwe, owing both to his status as a missionary and his background as a Mississauga Ojibwe, from a different and distinct community of Ojibwes in Canada.[4] Despite a shared language, he clearly was not entirely at home among these western Ojibwe communities, and the structure of his book comes to reflect a sense of radical unmooring as he leaves Rice Lake, returns, and then sets off again. The shape of these journeys divides *Life, History, and Travels* into three sections: (1) a history of Copway's life as an unconverted Ojibwe, with significant foreshadowing of future peregrinations; (2) the travels of a Methodist minister in distant lands; and finally (3) a return home, with a new and dramatically different sense of self and world that begins to unfold afterward. Reading this text as a travel narrative offers an important window into Copway's historical experiences and his evolving sense of place and the use of literary writing as a resource for nineteenth-century life.[5] Ultimately, that surge of emotions in 1839, including the dark sense that "every sweet has it[s] bitter," can be taken as a haunting precursor to what scholars today call solastalgia, a psychoterratic illness in which individuals' "mental wellbeing (psyche) is threatened by the severing of 'healthy' links between themselves and their home/territory."[6] For Copway, homesickness reflects a sudden and dramatic break between nature and culture, between himself and the "home" and community that he knew as a child in Rice Lake. Even as he returns there, he can never quite be "home" again.

Copway and the Ojibwe Nation

When *Life, History, and Travels* hit literary markets in 1847, it became a success almost overnight. Published under several titles on both sides of the Atlantic, Copway's autobiography went through seven printings by 1848, and by 1851 he'd written two revised versions, as well as a tribal history of the Ojibwe, *The Traditional History and Characteristic Sketches of the Ojibway Nation* (1850), and *Running Sketches of Men and Places, in England, France, Germany, Belgium, and Scotland* (1851), an account of his travels in Europe, where he attended the 1850 International Peace Conference in Frankfurt. He also founded a short-lived weekly newspaper in New York, *Copway's American Indian*, and in 1850 he proposed a plan to the US Congress for the establishment of an independent Indian territory east of the Missouri River, *Organization of a New Indian Territory, East of the Missouri River* (1850).[7] During this time, Copway traveled throughout the northeastern United States, and in the process he interacted with writers such as James Fenimore Cooper, Henry Wadsworth Longfellow, and Washington Irving, as well as Henry Rowe Schoolcraft and Francis Parkman. Thoreau himself owned a copy of *Traditional History and Characteristic Sketches*, which he annotated heavily.[8]

Despite its wide circulation and influence and its significance as an early example of Native life writing, *Life, History, and Travels* remains understudied. As Cathy Rex writes, critics have tended to emphasize "the general downward trajectory of Copway's life." This scholarship focuses on "his constant (and often disreputable) self-reinventions and identity changes" and "his (perceived) failure to project any identity beyond that of the 'Noble Christian Savage.'"[9] What emerges is an image of Copway as stuck between two irreconcilable worlds, forced to play the part of "both systems of belief in his various public guises, one as Kahgegagahbowh, 'Chief of the Ojibwa Nation,' and the other as the Reverend George Copway, 'Missionary to his People.'"[10] Such an impossible balancing act, one critic writes, "forced him to undergo various personality transformations that finally broke his spirit," leading to a downward spiral from which he would never ultimately recover, socially, financially, or otherwise.[11] Another scholar has even gone so far as to dismiss Copway's writing, arguing that he "had begun by the summer of 1851 to lose all touch with reality," and that while his life as a "literary celebrity in the United States provides rich material suitable for a television documentary drama or stage play," his work cannot be trusted as a verifiable historical source.[12]

Copway's character was called into question during his own lifetime, too. In 1846, shortly before Copway began to write his autobiography, he faced charges of embezzlement and theft from the Saugeen Mission and his elders at Rice Lake. Unable to pay these debts, Copway was sent to prison at Toronto, where he remained for several weeks before the Indian Department finally dropped charges, unwilling to incur legal expenses and doubtful of recovering the money.[13] These details are, perhaps unsurprisingly, not accounted for in his autobiographical writing, though they loom large in scholarly assessments of his work. And indeed, this moment marks the end of Copway's close relationship to Ojibwe communities, whether in Rice Lake or in the United States. Still, by largely dismissing Copway entirely, much of the scholarship also undermines any sense of an Ojibwe future that might emerge from *Life, History, and Travels*.

By presenting Copway as "stuck" or disreputable, this scholarship perpetuates what Maureen Konkle describes as "one of the most prevalent critical assertions about Native peoples and their historical relations with EuroAmericans"—"that EuroAmericans gave Native peoples a 'choice' to 'assimilate or become extinct.' "[14] But for writers such as Copway, the nineteenth century presented a far messier reality. Rather than focusing on what Robert Allen Warrior has called "parochial questions of identity and authenticity," we can more productively read *Life, History, and Travels* in terms of the future that Copway attempts to negotiate for himself and his community within this emergent political landscape.[15] To most productively read his work, we must turn specifically to the *travels* of Kah-ge-ga-gah-bowh, to the ways in which he was, in fact, radically unstuck in a world that seemed suddenly much larger and more dangerous, but also filled with possibility.

As Richard Scott Lyons puts it, "If anything can be considered an enduring value for Ojibwe people, it has got to be migration."[16] According to Lyons, the first treaties between the Ojibwe and the United States marked a pivotal moment in the tribe's evolving relationship with colonial powers, other tribal groups, and even themselves. Until the eighteenth century, he writes, *Mooningwanekaaning*, or Madeline Island, had been the center of tribal power. In Ojibwe legend and oral history, ancestors of the Ojibwe traveled to the Great Lakes from the eastern seaboard, along with the Odawa and Potawatomi, who split off along the journey and settled further south and east. Collectively, these three groups remained allied through the Council of Three Fires.[17] The Ojibwe, continuing westward, eventually came to a turtle-shaped island off of the southern shore of Lake Superior, where, as foretold in the prophesies, food grows on the water—*manoomin*, or wild rice.

Over the course of the eighteenth century, the economic system initiated by the fur trade had sent small groups of Ojibwe further south and west than ever before, and alliances changed as power became increasingly contingent upon trade connections and alliances with European nations. Among the suddenly dispersed communities of Ojibwe, Lyons writes, "any political ties that may have once existed had long since atrophied."[18] European and American power in the region was shifting, too. At the conclusion of the French and Indian War in 1762, the entire Ojibwe territory, once part of New France, came under British rule. After the American Revolutionary War, a large portion of the territory was ceded to the United States, though the British continued to occupy military forts south of the present international border into the early nineteenth century.

Only after the establishment of the Michigan Territory in 1805 and then the War of 1812 would the United States take full control of the region and begin entering into negotiations with the dispersed bands of Ojibwe.[19] Then, treaties at La Pointe, Fond du Lac, and elsewhere "characterized these groups as a single political entity, and since treaties are by definition contracts between nations, it turned them into a 'nation.' "[20] Once scattered and decentralized by migrations in search of furbearing animals, treaties forced Ojibwe to reorganize yet again.

Konkle writes that the phrase "Indian nation" entered Anglo-American legal discourse in the mid-eighteenth century, picking up significant weight when the United States attempted to legitimize its claims to territory through legal agreements with tribal groups. This notion of a new political autonomy among the tribes, of Indian "nationhood," has often been understood as "not really 'Indian' because it is a product of colonization and settlement."[21] As Konkle argues, however, such a position denies Indigenous peoples the experience of time and modernity. Lyons is even more resolute on this point. As he puts it, "the modernization that was initiated by treaty signers" is "an unfinished project that can and should be pursued further."[22] He argues that nineteenth-century treaties and the signatures on them—x-marks—be understood as a means of projecting Ojibwe political autonomy and self-determination into the future. Copway would later become involved with these treaties directly, even as he expressed doubts about the faithfulness of the United States. Indeed, the conclusion of *Life, History, and Travels* is predominantly an account of political organization among the Ojibwe in both the United States and Canada.

From 1834–7 and again from 1840–2, Copway traveled in the present-day states of Michigan, Wisconsin, and Minnesota. Although he does not mention them explicitly, he was almost certainly aware of two major land treaties affecting Ojibwe in the United States during his first years as a missionary, the Treaty of 1836 and the Treaty of 1837, which together ceded roughly 25 million acres to the United States in exchange for cash payments and usufructuary rights to continue hunting, fishing, and gathering within the region. These agreements made dramatic changes to communities where Copway worked, including Ottawa Lake (Lac Courte Oreilles) and what he calls Aunce (today L'Anse, near Keweenaw Bay). Following his 1839 return to Rice Lake, Copway witnessed a third agreement between the Ojibwe and the government of the United States. In the wake of this agreement, the Ojibwe Nation would be forced to reorganize once again. The Treaty of 1842 extended the ceded territory into the Lake Superior watershed in Wisconsin and the western Upper Peninsula of Michigan.

Interestingly, Copway is listed as a claimant in the agreement, to be awarded $61.67 for his debts as a missionary within the territory.[23] Besides marking what Copway will call the "ruin" of this region, the treaty also marks a veer in the text. Shortly after the treaty was signed, Copway left the United States, perhaps in part because he sensed the coming battle over removal.[24] Within the text, however, his decision to leave this part of "our nation" is simply a professional and personal choice. As he relates, he left only after receiving letters from missionaries in Canada "requesting me to return home and labor with them" (*Life* 143).

After the Treaty of 1842, the narrative arc of *Life, History, and Travels* is essentially closed. With this movement completed, the story comes to a halt, and Copway concludes by describing the General Councils of the Christianized Ojibwe at Saugeen and then offering a sketch of the broader Ojibwe Nation. The ultimate direction of *Life, History, and Travels,* then, is toward a specific kind of what Gerald Vizenor termed survivance—not of one individual, but of a modern, self-governing (and Christianized) Ojibwe Nation.[25] More specifically, the movement of the narrative is from an individual childhood at Rice Lake, the personal and familiar history of Copway himself, to this moment of nation-building. By setting out from Rice Lake and making his way through the larger Ojibwe Nation, Copway's travels are not merely a geographical tour, but a deviation from the lifeworld of a single Ojibwe community (and community member) into a larger, more nebulous political entity that was, at least in 1847, still taking shape. Moreover, by traveling to the United States at precisely this historical juncture, Copway stepped into what we might think of as a massive reorganization of dynamic, well-established communities—a veritable whirlpool of (former) French trade, British colonial rule, and American treaties; of traditional Ojibwe beliefs, Jesuit missionaries, and Protestant conversions; and of diverse, subdivided groups entering new political alliances as a means of claiming agency. *Life, History, and Travels*, then, attempts to navigate the dramatic reorganization of communities and lands that was taking place while he traveled and wrote—and through all of it, Copway attempts to maintain an image of home.

Affect and Mobility in *Life, History, and Travels*

Copway's autobiography can be divided into three distinct sections, each demarcated by a new relationship with his home community of Rice Lake. The first of these sections, spanning chapters one through eight, recounts Copway's life up to the summer of 1830, when he converted to Christianity at eleven years old. Throughout these chapters, Copway writes as an historian and an ethnographer:

> I have not the happiness of being able to refer to written records in narrating the history of my forefathers; but I can reveal to the world what has long been laid up in memory; so that when I "go the way of all the earth," the crooked and singular paths which I have made in the world, may not only be a warning to others, but may inspire them with a trust in God. And not only a warning and a trust, but also that the world may learn there once lived such a man as Kah-ge-ga-gah-bowh, when they read his griefs and his joys. (*Life* 70–1)

By thus framing his life and the text before us, Copway emphasizes his own liminal position between predominantly oral and emerging print cultures, between an Ojibwe past and an emergent, literary future. Narratives like his, he suggests, are one means of navigating historical change and, more importantly, of Indigenous agency. Furthermore, in these chapters, home comes to figure as Copway's immediate physical and cultural environment, specifically the region claimed by his ancestors, including his great-grandfather, "the first who ventured to settle at Rice Lake, after the Ojebwa nation defeated the Hurons, who once inhabited all the lakes in Western Canada" (*Life* 71). In this section of the text, Copway's home is figured as a traditional and familial territory, the places in which he has come to understand himself as an Ojibwe youth within a wider world.

The second section begins in chapter nine, when Copway converts to Christianity, attends the mission school at Rice Lake, and is selected by Methodist missionaries to travel west to Lake Superior. This section marks a dramatic turning point in the text, and also a slightly different emphasis on affect and sentiment. Here, home remains the immediate territory in and around Rice Lake, though not without discomfort. Over the course of chapters nine through twelve, the narrative tracks the peregrinations of a traveling missionary throughout the Northwest, first from Rice Lake to Sault Sainte Marie and La Pointe and finally to Illinois. While the history of his early life includes a few vivid anecdotes, this section turns forcefully to Copway's present moment, to the often-unpleasant feelings of nineteenth-century life in the Ojibwe Nation, from homesickness and isolation to physical hardship and racial discrimination. It also marks the beginning of what seems like an excursion, a journey from his childhood home out into a wider, quite surprising world. Over the next five years, he would experience a series of unanticipated joys and hardships, including a canoe voyage that he calls "the hardest journey that I ever made" (*Life* 115). This section culminates in two years of formal education at the Ebenezer Manual Labor School, followed by a whirlwind tour of the northeastern United States, with visits to New York as well as Boston. Throughout, home remains centered around Rice Lake, and specifically the familial connections and place-based identity that Copway knew as a young person there.

The third section begins in the late autumn of 1839, following that trip through New York and New England, when Copway finally casts his gaze upon Rice Lake after more than five years away. At this point in the text, Copway has come full circle, and his narrative voice, now fully embodying his role as missionary, has matured. Having closed this narrative arc, Copway's sense of homesickness seems to change, as well as his relationship with Rice Lake more broadly. In chapters thirteen and fourteen, Copway briefly mentions his 1840 marriage to Elizabeth Howell, apologizing for including such personal details and yet insisting on "speaking of one who has sacrificed so much in becomeing [*sic*] the partner of an Indian missionary" (*Life* 128). He is then elected by the community of Rice Lake before returning to the West as a missionary. In November, he arrives at the Elk River Mission, where he experiences conflicts with the Santee Sioux, eventually leaving for St. Peters, at present-day Minneapolis. He then establishes a new Methodist mission at Fond du Lac, describes a "Council of the Ojebwa Nation" and the subsequent Treaty of 1842, and then leaves for Canada again, perhaps because of the looming threat of removal. He begins serving as missionary at Saugeen in 1843, a mere three hundred miles from Rice Lake. By this point in the text, Copway has fully adopted the identity of an Indian missionary, and the sense of homesickness, indeed, the very idea of "home" itself, seems to change again. Indeed, after 1839, his family figures only in a "Geographic Description of the Ojebwa Nation," the final chapter of the text, tacked on almost as an appendix for the aid of understanding the current political situation he is navigating. By the time of his appointment to the Saugeen Mission, Copway's sense of home seems to have become centered around the larger Ojibwe Nation.

By considering the text of *Life, History, and Travels* in these three sections, readers come to see Copway's shifting emphasis on home and mobility. Departures, journeys, and homecomings structure the entire narrative, and in each section, the meanings of "home" and "away" take on slightly different significances. Indeed, though he first traveled west in 1834, even the earliest chapters of his narrative foreshadow a life marked by an uncommon degree of mobility. For instance, in recounting his early life before converting to Christianity, Copway describes a dream he experienced as a twelve-year-old boy. While fasting and seeking visions, he writes, "we were to leave the wigwam early in the morning, and travel all day from one place to another, in search of the favor of the gods" (*Life* 87). During one such fast, Copway is sleeping away from his wigwam with others, presumably people of his own age, when a person approaches from the east, "walking on air." This person addresses him directly and in a striking fashion—"Is this where you are?"—and Copway offers a simple "yes." Then the stranger instructs him to observe a tall pine tree, which Copway notes as "lofty, reaching toward the heavens," with "branches extended overland [*sic*] and water," and roots "very deep" (*Life* 87). This strange figure then begins to sing and, as Copway recalls, the tree "commenced waving its top; the earth about its roots was heaved up, and the waters roared and tossed from one side of their bed to the other" (*Life* 87). When he ceases singing, everything is once again still.

The power of this strange figure, whom Copway's father calls "the god of the winds," is immediately apparent, but his interest in the young man remains unexplained. Moreover, the odd, provocative greeting-by-way-of-question—"Is this where you are?"—remains unsettling. The figure goes on to offer a lesson:

> "Now," said he, "sing the words which I have sung." I commenced as follows:—
>
> "It is I who travel in the winds,
> It is I who whisper in the breeze,
> I shake the trees.
> I shake the earth,
> I trouble the waters on every land."
>
> While singing, I heard the winds whistle, saw the tree waving its top, the earth heaving, heard the waters roaring, because they were all troubled and agitated. Then said he, "I am from the rising of the sun, I will come and see you again. You will not see me often, but you will hear me speak." (*Life* 87–8)

Immediately after speaking, the figure, who Copway now refers to as "the spirit," turns and departs, apparently having blessed Copway with a power of his own. The fundamental meaning of this dream seems straightforward enough, and in the lines that immediately follow Copway offers an interpretation in the words of his father, who posits that "the aged tree, I hope, may indicate long life; the wind may indicate that you will travel much; the water which you saw, and the winds, will carry your canoe safely through the waves" (*Life* 88).

In relaying this dream and offering his sense of what it meant for him even as a child, Copway must navigate the tenuous ground between, on the one hand, his audiences' intense interest in "authentic" Ojibwe culture and, on the other hand, his new role as an intellectual and a missionary. Copway's language in relating the significance of his dream, from his use of the third-person objective stance to the insistent use of the past tense, suggests a sense of doubt over the interpretation of these "omens," though he will not quite dismiss the meaning of his visit from the spirit of the wind. To thread this tight space, he will offer a seeming wink to the reader. "But, however, little reliance can be placed in dreams," Copway writes, "yet may not the Great Spirit take this method, sometimes, to bring about some good result?" (*Life* 88). With this question, Copway's dream, though it is presented as part of his history of a pre-Christian life, remains available, however quietly, for the reader's interpretation. Here then, is a seed not of doubt, but of possibility. Might not the Great Spirit, or God, sometimes offer information in a dream? Might not a nineteenth-century Ojibwe traveler, making his way west toward *Mooningwanekaaning*, rely on dreams just as his ancestors during the Great Migration? By gesturing toward such questions, even in this indirect way, the narrative of *Life, History, and Travels* establishes a connection, however slight, between traditional Ojibwe narratives and his current, modern literary form. Moreover, he establishes travel and mobility as a grounding dimension of his identity as a Christian missionary and a Mississauga Ojibwe. This dream, though Copway might seem to dismiss it, thus emphasizes the importance of travel even in this first section of the book, wherein "home" is straightforwardly the environment about Rice Lake.

Throughout its second section, *Life, History, and Travels* establishes clear connections among Christianity, mobility, and writing. Copway converted to Methodism in the summer of 1830, a moment marked as much by finding his (Anglophone) literary voice as by a sense of Protestant religious awakening.[26] Up until that summer, he'd "only begun to spell and read," but with a newfound passion for understanding and sharing Christian scriptures, he resumed his studies "with a new and different relish" (*Life* 102). This moment is marked, too, by an emphasis on connection to family and to community. By coming into Christianity, Copway is remembering his pious mother, who died only months before, and becoming closer to his father, who himself converted to Methodism in 1827.

In the moments before his conversion, Copway describes being physically ill, writing with emphasis, "I now began to feel as if I should die; *I felt very sick in my heart.* Never had I felt so before; I was deeply distressed, and knew not the cause" (*Life* 101). After fainting for some unknown period of time, he revives and finds his father, who embraces him. In the moments that follow, the physical space around him comes to appear different, charged with a positive energy:

> As I looked at the trees, the hills, and the vallies [*sic*], O how beautiful they appeared! I looked upon them, as it were, with new eyes and new thoughts. Amidst the smiles of creation, the birds sang sweetly, as they flew from tree to tree. (*Life* 102)

Earlier in the text, Copway's physical environment is portrayed as a familial home inherited from his ancestors, and Copway himself as "one of Nature's children" (*Life* 73). Here, however, his observations of the environment take on a different character. Writing in the language of Protestant conversion, but still with a distinctly Ojibwe perspective, Copway's sense of home becomes suddenly grounded in these "emotions of pleasure."

Many scholars have read Copway's autobiography in terms of Protestant conversion narratives, as well as contemporary writing by his fellow Methodist missionary, William Apess.[27] To be sure, the form of Copway's writing in this chapter resembles that of John Bunyan's *Grace Abounding* (1666) and *Pilgrim's Progress* (1678). As Barry O'Connell notes, these narrative forms were deeply ingrained in Protestant religious communities, often read aloud in households with few other books except a Bible.[28] These texts, too, emphasize the emotional response to religious experience. For Copway, though, conversion is somewhat different. Religion in this passage is above all else a means of connecting to family, and also of joining other Ojibwe in a community being actively reorganized in the face of colonization. It is, furthermore, a means of finding his literary voice, not to mention a recognizable form with which to relate and understand the events of his life. Following his conversion, Copway describes a happy period at home in Rice Lake, enjoying regular contact with his father and other relatives, even to his departed mother, who he now remembers with particular joy. For the next four years he regularly attended his father's class meeting, "and as often as possible I attended school" (*Life* 103). Then, quite suddenly, Copway describes an unexpected opportunity to leave Rice Lake.

In 1834, Copway relates, Reverend John Clark, the Superintendent of the missions on Lake Superior, "requested that two native preachers and two native teachers" be sent to assist in ongoing work in the heart of the Ojibwe homeland (*Life* 103). Alluding to his success in school and prayers "that God might send me to instruct the children in the truths of religion," Copway is told to accompany his uncle, John Johnson, along with the preachers John Taunchey and John Caubage, to the south shore of Lake Superior. The decision to go remains in his hands, however—and the dramatic consequences of this choice lend Copway's prose a teetering, back and forth effect, not to mention a drawn-out sense of discomfort:

> One day I determined to leave the village so as to avoid going to Lake Superior; I hunted along the river Trent, hoping that John Taunchey would be gone before my return; I felt very unwilling to go. I was absent over two weeks; they were the longest two weeks I had ever experienced. Yet the whole time I felt dissatisfied; something seemed to whisper to me, "George, go home, and go to Lake Superior with your uncle John Taunchey." I returned to the village. The first person I saw, informed me that my uncle was waiting for me, and that my father had left it to me to decide whether to go or to stay. Here I was; the missionaries came, and said, "George, your father has left it to you to go or stay. It is your duty to go; John is waiting, and to-day you must conclude." Our school mistress, Miss Pinney, came and reasoned with me. I recollected, too, that I had prayed that God might prepare me to be useful to my brethren; and now, that I had some good reason to think that my prayers had been heard, and still to refuse to go, would perhaps be acting in opposition to the indications of God. I wept and prayed; but O! that night of struggle! I could not sleep. In the morning, I said to my father, "I have concluded to go; prepare me for my journey." (*Life* 103)

This passage is the single most dramatic veer in *Life, History, and Travels*, even more significant and pronounced than Copway's conversion to Christianity. Following this decision, "home" becomes a *remembered* place of connection. His decision to leave marks the sudden conclusion of a youth at Rice Lake and the beginning of Copway's subsequent identity as a footloose missionary, never remaining in any one place for a significant period of time, but frequently looking back longingly toward home. His departure, in other words, is a radical unmooring, and Copway's identification with his "brethren" reflects a new sense of place. As a missionary, he is forced to reimagine how he fits within a much larger world. Certainly, the sense of "duty" invoked here implies a larger benefit to other Indigenous peoples in North America, and perhaps also to white readers, but it also clearly signals what is already an emerging sense of a separation from his former way of living and being in the world, initiated by his conversion and crystalizing here. By succeeding in school, he suggests, he has set himself up to leave Rice Lake. Indeed, Copway's separation from his family is never resolved within the text—for the rest of *Life, History, and Travels*, his father is mentioned only twice, once in a homesick reverie, and again in a list of significant chiefs.[29]

I'm interested here in the form of this text, the shape and conventions of Copway's prose before and after his initial departure. Hertha D. Sweet Wong, writing on the importance of *Life, History, and Travels* in establishing a model for later Native authors, notes that the book uses the narrative structure of a Christian conversion to produce something altogether different, "an interweaving of myth, history, and contemporary incidents as well as an intertwining of personal experience with tribal history and culture."[30] To be sure, Copway does draw from various literary conventions in shaping and framing the narrative, but the degree of mobility, indeed the *travels* indicated in the book's title, suggest the importance of travel writing. The book meanders across genres and traditions, but ultimately it takes the shape of an excursion, with a distinctive turn back toward home followed by another departure. William Bevis describes a certain "homing instinct" as a key feature of Native American novels, in which home emerges "as a primary mode of knowledge and a primary good."[31] In this way, Rice Lake functions as the north star throughout this section of *Life, History, and Travels*, and perhaps even after Copway learns that he can never "return" in quite the same way. Many nineteenth-century travel narratives describe a sense of being "unsettled" by the wider world, having witnessed communities and places in the midst of incredible sociocultural and environmental change. Often, however, these writers solidify their journeys in terms of an adventure or a vacation, thus reorienting that wider, more threatening world to a relatively stable sense of home in New England, New York, London, or elsewhere. *Life, History, and Travels* is unique because it must reorient not simply the journey or the destination, but, more fundamentally, Copway's sense of home and of himself. To ignore the primary function of travel writing, in short, is to overlook the most significant turning points in *Life, History, and Travels* and the degree to which the text rewrites Copway's sense of place.

After he decides to leave Rice Lake, Copway's expression of his relationship to home immediately changes. From this point on, the narrative is suddenly and definitively a travelogue, recounting new and novel experiences as it whirls around a region undergoing rapid cultural, economic, and ecological change in the face of colonization.[32] It is difficult for twenty-first-century readers even to keep up, as Copway's place names, anecdotes, and digressions bear witness to the shifting histories of Ojibwe, Dakota, French, British, and American dominance. In thus describing these travels, Copway attempts to place himself within a new and unsettling but also enchanting global geography. The pleasant affects associated with his conversion are immediately changed, too. From this point until the fall of 1839, when Copway finally returns to Rice Lake, the text bears witness to an intense, simmering sensation of homesickness, a feeling that rises to the surface of the text in moments of stress—and such moments are everywhere.

From Rice Lake, Copway travels to Sault Sainte Marie, arriving on or about August 24. It is here, some 300 miles and several weeks after his departure, that Copway's homesickness first emerges within the text. Chapter ten opens with the first description of this new affect:

> I now began to feel the responsibilities resting upon me. The thought of assuming the station of a teacher of the Indians, with so few capabilities, was enough to discourage more gifted men than myself. Frequently did I enter the woods and pour out my soul to God, in agony and tears. I trembled at what was before me; and said, "who is able for these things?" (*Life* 105)

This passage concludes with a "still small voice" which answers, "My grace is sufficient for you" (*Life* 105). Copway's homesickness, though, remains unabated. Indeed, it only seems to grow as the narrative progresses.

When Copway leaves Sault Sainte Marie he travels west, accompanied by a white missionary, Rev. Chandler, as well as his uncle John Taunchey and a group of "traders who intended to winter on the shores of Lake Superior and do business with the Ojebwas" (*Life* 105). Copway calls these French traders "the most wicked of men," who "would gnash their teeth at each other, curse, swear, and fight among themselves," making an already difficult journey all the more unbearable (*Life* 105). Over these three weeks on Lake Superior, during which they spend an entire week "weather-bound," waiting out a period of storms and rough swells, Copway's homesickness comes to take on a new dimension:

> Every Sabbath I devoted about an hour in sighing and crying after *home.* What good can I *do*, when I reach the place of labor? was a question that often occurred to my mind. Still, we were going farther and farther from home. (*Life* 105)

In ritualizing his sentimental turns toward home, Copway signals that these sensations will never quite vanish. Indeed, they seem to continue growing, and they remain present even when he does not mention them explicitly. Furthermore, this process of *imagining* home, of calling upon visions of Rice Lake during particular trying moments, will come to be an important means of navigating the experiences ahead of him.

That winter, Copway and his companions work in a region that he calls "Aunce Bay," near the present-day communities of Keweenaw Bay and L'Anse, Michigan. Their mission, he writes, is "no better than a wigwam," but it must serve "as a dwelling, a school house, a meeting house, and a council room" (*Life* 105). Despite settling here for the season, Copway and his companions continue to move about the region, often traveling on snowshoes as they "visited the Indians daily, for the purposes of praying and conversing with them" (*Life* 106). "Each day," Copway writes, invoking a travel metaphor, "we took a different direction in visiting the unconverted. We would sing, read the scriptures, and then pray with them" (*Life* 105). By noting the continued mobility of this winter's work, Copway underscores the fact that though "settled" for the winter at this mission, he remains a visitor among these Indians, a traveler in a foreign land. He highlights this again in describing the joy of a Sabbath meeting, during which "every converted Indian would try to induce his relatives to embrace religion" (*Life* 106). "These happy scenes," he writes, "often made me forget home" (*Life* 106). "Often," the reader might suspect, but seemingly not *never*, and certainly not *for good*.

In the spring, Copway and Taunchey are relieved of their positions near Keweenaw Bay, and Copway records their memorably smooth journey back to Sault Sainte Marie with Superintendent Clark, raising a sail in a 36-foot bark canoe and, with a favorable wind, making record time. "The traders were much surprised," writes Copway. "The Indians could hardly think it possible for any person to travel the distance in so short a time" (*Life* 109). After a few weeks, they set off again for Ottawa Lake in the west, near the present-day community of Lac Courte Oreilles. This time, however, they experience rough seas and fog, running out of provisions after spending several days wind-bound on the beach. Just east of La Pointe, exhausted, starving, and low in spirits, Copway records his most poignant instance of homesickness.

It is late on a Saturday evening, and Copway's party is adrift on Lake Superior, within sight of shore. They have been toiling for days, and Copway is seated in the stern of another bark canoe, attempting to steer, when his companions' paddles go still. The entire group has fallen asleep, and Copway cannot wake them, cannot even force himself to remain awake. He does his best to steer for the shore, where he wraps himself in a blanket and collapses on the sand. The next morning offers little relief:

> Sabbath morning came. I had dreamed that we were just about sitting down to a warm breakfast, when Peter Marksman woke me, and said, "George, come,

> get up, *blackfast*" (breakfast, he meant, he could speak but little English)[.] If it had not been the Sabbath, I might have been induced to retaliate. It was, indeed, a *blackfast*, dark enough; nothing to eat, and only tea to drink for breakfast, dinner, and supper! (*Life* 111)

They have but fifteen miles to go; indeed, they can *see* La Pointe in the distance, but the party will not travel on a Sunday—and indeed, this religious practice offers little comfort.

Hungry and exhausted here on the shores of Lake Superior, Copway turns back to the ritual of *imagining* home, of recreating the Rice Lake that his readers know, by now, is always sifting about in Copway's memories, if not always on the surface of the text. Copway writes:

> I walked in the woods, and all that I could think of while reading my Bible, was *home*. I looked toward *home*, and wept at the thought of it. I said to myself, O my father, if you knew my situation to-day, you would feel for me, and fly, if possible, to assist me! I feel that your prayers ascend for me; and then descend like gentle rains, into my soul. *Home! home!* however humble, it is still *home*. (*Life* 111–12)

This passage is striking for its use of the language of conversion, including even the "pleasant emotions" that Copway previously associated with connection to his family and a community of Christian-Ojibwe around Rice Lake. Moreover, the repetition and emphasis on home place an uncommon weight on these Sabbath reflections, a ritualistic remembering of home, of a former life. To continue depicting these low points, to underscore them again and again, would be to undermine his status as a missionary and intellectual. To portray himself in such a state here, however, and to conflate homesickness and "my father," and thus family and God, Copway frames his travels in terms of the spiritual journey toward the work that awaits him.

Copway's difficult journeys are far from over, though. The following morning, well before dawn, the party is on their way to La Pointe where they finally find their breakfast and resupply. From there they make their way inland, struggling over long portages to reach a remote community where he will work to establish a new mission. In early October he arrives at Ottawa Lake, where he spends the winter laboring in a "day-school" with some "few scholars" (*Life* 113). At one point, Copway faces a famine. Out of food and unable to hunt or fish in the late winter, he is faced with a difficult journey back to Lake Superior, some 170 miles over difficult terrain, urged on by hunger. Copway recalls blistered feet from the straps of his snowshoes. "The young man who accompanied me," he adds, "suffered still more, for the blood was oozing out of his moccasins" (*Life* 113).

Copway returns to that mission again the next winter, where there are "but few Indians," and yet still "too many for one teacher." His fellow missionaries expected him to travel to La Pointe in the late fall of 1836, but Copway recalls a certain "sense of duty" to "spend the winter and spring in teaching, singing, and praying among the people" at Ottawa Lake (*Life* 116). He remains there until the spring of 1837, when he sets off to the south, eventually making his way to the Ebenezer Manual Labor School in Illinois. Through all these travels, the reader comes to assume, he occasionally turns his gaze back toward home, toward the sense of connection that he felt in Rice Lake.

When Copway finally returns to Rice Lake in the late fall of 1839, he is no longer the same youthful Christian-Ojibwe who stood on those shores in 1834, preparing to "bid farewell to all we loved so tenderly." In one sense, these travels have changed him, and the community of Rice Lake has changed in his absence. Many of his relatives have "left this, for a better life," in his words (*Life* 125). Moreover, he comes to see Rice Lake differently. After returning home in 1839 and throughout the third section of the book, Copway makes regular, explicit references to treaties and political organization. Copway will continue to use the discourse of travel writing, wending his way west again and then north to familiar sites along the south shore of Lake Superior. By the spring of 1840, however, these peregrinations are no longer oriented toward Rice Lake and the "life" and "history" of Copway. Rather, *Life, History, and Travels* attempts to mediate the growing tensions between Ojibwe communities, colonial powers, and Protestant missions.

The book's first detailed description of Ojibwe political organization occurs immediately after Copway's return to Rice Lake in 1839. Early in 1840, Copway participated in the General Council of the Nation, a gathering of Ojibwe leaders from communities in Canada. The council had submitted a petition to the Canadian government, which Copway appends to the text, suggesting relocating the "Credit Reserve" to "a good tract of land" more suitable for agriculture and away from the influence of settlers (*Life* 135). The reply from Charles Poulett Thomson, the first Governor General of Canada, Copway writes, offered "but little satisfaction": "His Lordship did not even deign to affix his *name* to the note. Since then, nothing has been heard of our papers, and therefore we must conclude that they have been laid *under* the table" (*Life* 127). This first, decidedly *bitter* taste of political negotiations marks an important turning point in the text.

From this point through the remainder of the text, Copway's emphasis suddenly shifts from the meaning of his own journeys to, more broadly, the political future of Ojibwe communities.

In 1840 Copway makes his way west to Elk River, in the present-day state of Minnesota, where he spends a winter among an Ojibwe community amidst an ongoing conflict with "their enemies, the Sioux, whom they dreaded" (*Life* 133). This region is "a dangerous place for the Ojebwa missionaries," he writes, "but not so for the whites" (*Life* 138). Copway goes on to propose that "Christianity and education alone" can resolve this very old conflict, but his proposed solutions also seem to reflect the growing number of missions in the region. Among his seven proposed conclusions, three suggest various forms of cooperation between Protestant missions of different denominations, from limiting "interference" to assisting each other and, finally, to not preaching "their own peculiar doctrines, to the disadvantage of other denominations" (*Life* 141). These suggestions, besides foregrounding Copway's role as an intermediary figure among Ojibwe communities and missionaries, hint at the power and also the limits of missionary work. The sudden mention of other denominations further suggests Copway's controversial belief that the overall work of missions transcends conflicts between specific faith communities.

In the spring of 1841, Copway travels to Fond du Lac, at the head of Lake Superior, to establish a new mission. Here, the text again considers the relationship between Ojibwe communities and colonial powers. In 1842, still living at Fond du Lac, he travels to La Pointe to observe treaties between the Ojibwe and the United States:

> The Council of the Ojebwa nation assembled in this place about the first of October. The government agent, R. Stewart, of Detroit, treated with them for their mineral regions, for which the government gave them a large sum of money. From this time, I shall date the dissipation, misery, and ruin, of this part of our nation. (*Life* 142)

The final line of this passage leaves little room for interpretation. Copway goes on to elaborate his position on the Treaty of 1842, but his sense of the damage initiated by this treaty needs little explanation. The Treaty of 1842 extended the ceded territory into the Lake Superior watershed in Wisconsin and the western Upper Peninsula of Michigan.[33] Interestingly, Copway is listed as a claimant in the agreement, to be awarded $61.67 for his debts as a missionary within the territory.[34] Copway left the United States soon after, perhaps in part because he sensed the coming battle over removal. The region was, indeed, dangerous for an Ojibwe missionary.

After the Treaty of 1842, the narrative arc of *Life, History, and Travels* is essentially closed. Returning to Canada a second time, the text records only a list of official positions, first at the Credit River and Saugeen Missions and then with the Indian Department. Copway briefly describes an intense moment when his young son, "about three years old," fell overboard a schooner en route to Saugeen Mission from Toronto, during which he jumps in to rescue the child (*Life* 144–5). The moment is fast-paced and vivid, but surprisingly brief. Then, the narrative moves on. As Copway writes, "We now resumed our labors at the [Saugeen] Mission" (*Life* 145). The significance of this moment, of these journeys, becomes secondary to the larger fate of the Ojibwe Nation.

With this movement completed, the story comes to a halt, and Copway concludes by describing the General Councils of the Christianized Ojibwe at Saugeen and then offering a sketch of the broader Ojibwe Nation. The ultimate direction of *Life, History, and Travels*, then, is toward a specific kind of what Gerald Vizenor terms survivance, or the manifest ways in which Native peoples rise above colonial simulations of Indian identity. Survivance, as Vizenor writes, can take multiple forms, from "the silence of heard stories, or the imagination of oral literature in translation" to "an aesthetic restoration of trickster hermeneutics."[35] By adapting the rhetoric of travel writing to a sense of Ojibwe migration, mobility, and adaptation, Copway records another form of survivance—not of one individual, but of a modern, self-governing (and Christianized) Ojibwe Nation.

Throughout his difficult journeys to the west, Copway has witnessed a world reeling from a radical new cultural and ecological instability, conditions that we today associate with the Anthropocene.[36] He has been at La Pointe during the brief rise of the American Fur Company, when local trade networks were still dominated by families of mixed Ojibwe, Metis, and French descent, and each clan and family still claimed traditional hunting and gathering grounds inland, where they lived seasonally during the winter hunting and trapping seasons, during the late winter maple sugaring season, and in the late summer, when *manoomin* was harvested and processed throughout the region. Traveling south to the Chippewa River, he has observed a mill where white loggers were cutting pines, the beginning of the Great Cutover, which within scarcely a generation would leave the entire region free of old growth white pines. Copway describes his own conversion as a kind of threshold moment, a liminal space that he witnessed countless other Ojibwe step across over the next few years. More broadly, we could describe his life as a liminal period in the Upper Great Lakes—the transition from the late fur trade to an era of logging and subsequent Euro-American settlement. Remarkably, Copway never despairs in these moments for the future of the Ojibwe. This orientation toward the future, toward better days that are, at the time, difficult to fully imagine, suggests something similar to what Jonathan Lear has identified in Plenty Coups, Chief of the Crow Nation, as "radical hope." This radical form of hopefulness is built around a conviction "that even with the death of the traditional forms of Crow subjectivity, the Crow can nevertheless survive—and flourish again."[37] Copway's memories of home, and also his experiences of homesickness, indicate the potency of this conviction, which extends beyond religious conversion and political observations to encompass something much more expansive. Homesickness, here, becomes a manifestation of adaptation and survivance.

Earlier in *Life, History, and Travels*, Copway writes of the sense of environment that he experienced as a child: "I was born in *nature's wide domain!* The trees were all that sheltered my infant limbs—the blue heavens all that covered me." He notes an "enduring love" of a personified nature and "emotions of pleasure that roll in my breast" in gratitude for this humble upbringing. "It is thought great to be born in palaces," he adds, "surrounded with wealth—but to be born in nature's wide domain is greater still!" (*Life* 73). In this sentimental appeal to the beauty of Creation, Copway hints at a Christian-Ojibwe environmental awareness, a unique sense of place melded to a new religious identity. Even here, though, in the beginning of the narrative, his prose is marked by an uncommon awareness of change. His memories, the reader senses, are tinged with an awareness of a world that is vanishing.

"I was born," Copway writes, "sometime in the fall of 1818, near the mouth of the river Trent, called in our language, Sah-ge-dah-we-ge-wah-noong" (*Life* 73). He goes on to describe the immediate setting that he remembers most clearly from his childhood:

> I remember the tall trees, and the dark woods—the swamps just by, where the little wren sang so melodiously after the going down of the sun in the west—the current of the broad river Trent—the skipping of the fish, and the noise of the rapids a little above. It was here I first saw the light; a little fallen down shelter, made of evergreens, and a few dead embers, the remains of the last fire that shed its genial warmth around, were all that marked the spot. When I last visited it, nothing but fur poles stuck in the ground, and they were leaning on account of

> decay. Is this dear spot, made green by the tears of memory, any less enticing and hallowed than the palaces where princes are born? (*Life* 73)

In this awareness of an environment in motion, Copway notes a strange, perhaps even strained, sense of permanence to natural features of the landscape. If this place is made "green by the tears of memory," it exists primarily in the imagination of the rememberer, not so much a real physical space as, rather, a specific *place* in the lifetime of Copway. Moreover, the comparison to palaces, while making an appeal to humble origins, takes a striking turn in the lines that follow. "Nature will be nature still, while palaces shall decay and fall into ruins," Copway continues. "Yes, Niagara will be Niagara a thousand years hence! the rainbow, a wreath over her brow, shall continue as long as the sun, and the flowing of the river!" (*Life* 73–4).

Because they are positioned relatively early in *Life, History, and Travels*, these memories of a childhood connection to place might at first seem like a nostalgic recollection of time spent at home, near Rice Lake. At the end of the book, however, when Copway's final return to home comes into focus, these lines take on a slightly different meaning. By the conclusion, when we turn away from Copway's personal sense of place toward a more general sense of a politically viable Ojibwe Nation, homesickness comes to represent a kind of proto-solastalgia, or the ultimate inaccessibility of one's home due to widespread political, social, and ecological changes. Although Copway remains physically present in this (threatened) cultural landscape, his sense of homesickness anticipates what Albrecht refers to as solastalgia, a "psychoterratic illness" in which one's home becomes inaccessible due to, for example, large-scale mining.[38] *Life, History, and Travels* is thus quite different from works by other writers examined in this book. If Copway's travels evince a dawning awareness of the looming changes coming to the south shore of Lake Superior and the Upper Mississippi watershed, these "still green" memories of home reflect a very different sense of the world around him.

For white writers, the disturbance of an excursion, the unsettling of previously held notions of one's world, comes most into view upon the return home, as when Fuller leaves Michilimackinac musing on the future of Indigenous peoples, or when Thoreau glances back at Cape Cod, vanishing in a silvery haze behind his steamboat. In *Life, History, and Travels*, on the other hand, the notion of an environment, or the space which encircles and environs a discrete observer, is centered not around a relatively stable sense of self, but rather a recollected sense of *home*. By the time he returns in 1839, Copway's Rice Lake, the home that he cries for in that trying moment east of La Pointe, represents a connection that he has lost, a way of living and being in the world that changed with his conversion and then strained and broke when he left home to labor among his "brethren," "the Indians." Copway's world, his environment, is indeed turning around him, a rapidly whirling milieu of Christian faith, Ojibwe cultures, and networks of travel, trade, power, and discourse. It remains centered, though, on the stable sense of "nature" as that world, that community, which exists most poignantly in his memories.

The form of *Life, History, and Travels*, the shape of Copway's prose as compared to other travelogues published by white travelers, reflects a structure of leaving home, looking back, and ultimately struggling to return. For him, returning to Rice Lake is not about physically traveling from the United States to Toronto and finally to his former home, but rather, about remembering the relatives, the connections, the enmeshment that he experienced as a child, and that remain an indelible source of personal strength in his writing. In all his wandering, all of his work at missions throughout the region, and all of his glances toward home, Copway strives to come to terms with his own sense of a diverse, newly congealed Ojibwe Nation on the brink of dramatic change. We must therefore read his travels not merely in terms of his political and professional maneuvering in the United States and Canada, but in terms of a deeply emotional effort to remain grounded, *placed*, in a world that continues whirling, that continues threatening Native peoples. While Thoreau would continue to develop and refine his mature excursion form, eventually questioning this stable sense of home in a dynamic, multicultural present, his writing never displays such dramatic veers as Copway's *Life, History, and Travels*. Indeed, some of Thoreau's most well-known travel writing constructs "nature" in terms of a destination, a site to explore. Only later, in one of his final travel narratives, would he confront the power of travel and travel literature to entirely reframe one's sense of self, home, and world. For Copway, meanwhile, travel writing offered above all else a means of adapting and bearing witness to a world that still includes Indigenous peoples, and that will still be "made green by the tears of memory."

5

Disorientation in *The Maine Woods*

> *Not until we are lost, in other worlds, not until we have lost the world, do we begin to find ourselves, and realize where we are and the infinite extent of our relations.*
>
> –Thoreau, *Walden*

Writing in his Journal in the late summer of 1857, Henry David Thoreau describes the moment when his friend and traveling companion, Edward Hoar, became lost in a particularly remote area of the Maine North Woods. This passage, which will serve as material for "The Allegash and East Branch," reveals a sense of panic unlike anywhere else in Thoreau's writing:

> I shoutd [*sic*] searched above & below this precipice ~~for half~~ in the twilight till I could not see—expecting nothing less than to find his body beneath—For half an hour I was afraid & believed only the worst—I thought what I should do the next day if I did not find him—what I *could* do in such a wilderness—~~who could help me~~—and ~~what~~ how his friends would ~~think~~ feel if I should return without him—[1]

Disorientation is clearly different here than in *Walden*, where being lost affords the opportunity to "realize where we are and the infinite extent of our relations" (*Wa* 171). This moment, by comparison, enacts a serious disturbance to Thoreau's way of being in and writing of the world.

While Thoreau searches frantically, his Penobscot guide Joe Polis walks down the river. Rather than shouting, he hoots occasionally like an owl, as he's done before along their journey—for this is not the first time that he's had to search for Thoreau and Hoar. Returning, he explains that there are tracks visible along the streambank, that Hoar can't be far ahead. Despite the fading light, Thoreau is "encouraged" to continue the search:

> This ~~relieved~~ encouraged me very much—I ~~suggested~~ proposed that we should both keep down the shore to the lake—but ~~he~~ The Ind. said ~~we could do nothing in the woods in the dark~~ "No use—can't do anything in the dark—come morning then we find him."[2]

Thoreau's revisions to this passage are illustrative of a profound anxiety. In the canceled line "we could do nothing in the woods in the dark," for which he substituted Polis's calm assessment of the conditions and the situation at hand ("No use—can't do anything in the dark—come morning then we find him"), readers witness an experienced, skillful guide at work. Having noted Hoar's tracks and insisting on the safest course of action, Polis will now attempt to help Thoreau see that his friend will be alright, will make his own camp and pass the night as comfortably as they will. Hoar has, after all, just returned from traveling in California, where he spent nights out in much rougher country. Thoreau, however, cannot fully appreciate Polis's assurances, his anxiety surfacing again in the final line recording the day's events: "Sometimes I doubted whether P. had really seen tracks—since he manifested an unwillingness to make much of a search—and then my anxiety returned."[3]

This unpublished Journal passage has not been widely studied by scholars of *The Maine Woods*, and the extent of Thoreau's revisions, the shock with which he records these events, is difficult to fully capture in conventional academic formatting. This rough, heavily revised text is important, however, because it illustrates the extent to which excursions are mediated by specific literary practices. The disturbances shaping Thoreau's impressions and ultimately his prose become meaningful only much later. The text thus reflects something like what Thoreau would describe to his friend H. G. O. Blake later that fall, a process whereby an excursion takes on its final significance only gradually, and only through revision, experience itself proving inadequate.[4] Throughout his three trips to Maine, Thoreau was challenged to imagine new ways of telling a story, veering away from adventure narratives, spiritual quests, and other forms to something new entirely. Ultimately, this night when Hoar became lost will mark a turning point in "Allegash and the East Branch," a moment when Thoreau and Hoar alike become more seasoned travelers, the party converging again around this experience and Polis's skilled leadership. In the aftermath, we come to see that Polis has been the leader of this trip all along, subtly teaching Thoreau to see and to navigate this landscape.

In *The Maine Woods* (1864), published two years after his death, Thoreau narrates three excursions to the North Woods of Maine. In the first essay,

"Ktaadn," Thoreau and his companions become decidedly lost in a dense forest on a remote mountainside, an experience which leads directly to a thrilling sense of "*Contact!*" with what he sees as vibrant, dynamic earthly matter, the raw material of both his own body and "some star's surface," the world itself (*MW* 71). The second, "Chesuncook," reports an 1853 moose-hunting trip, when Thoreau acted "as reporter or chaplain to the hunters," his cousin George Thatcher and Penobscot guide, Joseph Aitteon (*MW* 99). On their final evening in the North Woods, Thoreau sat before a fire with four Wabanaki men, mesmerized by their language and their stories.[5] His third essay narrates the 1857 trip, by far the longest and most strenuous, when he apprenticed himself with Polis and attempted, along with Hoar, to navigate the portages along what his maps called "Webster Stream."[6] Each of these excursions posed a specific literary challenge to Thoreau, requiring that he relinquish at least some degree of control and expectation in order to really see the "good purpose" of his journeys, as he put it to Blake.[7] The unsettling sense of being a little lost in a rapidly changing world runs throughout "Ktaadn" and "Chesuncook" before finally coming to a head in "The Allegash and East Branch," where "wilderness" travel becomes a matter of social (re)formation and (cross-cultural) group dynamics.

This chapter examines the most dramatic moments of disorientation in each of these essays to argue for a unified reading of *The Maine Woods* as a single, cohesive experiment in the excursion form, spanning Thoreau's years at Walden Pond and running to the end of his life and beyond, when his sister Sophia Thoreau ushered them to publication.[8] Unlike in his other books, *A Week on the Concord and Merrimack Rivers*, *Cape Cod*, and *Walden*, each of which compresses and combines experiences to form a single narrative, *The Maine Woods* enacts what Laura Dassow Walls has identified as a pattern of going back and revisiting, "a process catalyzed not by a single revelation but by the complications of returning, thinking, and then rethinking, original conclusions confuted, complicated, or expanded."[9] Together then, these excursions progressively unsettled Thoreau's methods of telling a story. By reading them together as separate but crucially connected excursions, we come to see the extent to which the concepts of "civilization" and "wilderness" and ultimately "home" and "away" become so dramatically unsettled through specific experiments in nonfiction storytelling. In *The Maine Woods*, foundational ideas about traveling and returning home are blurred by a new sense of the world itself, first as a dynamic, material planet and later as "home" for people and cultures of radically different ontologies.

"Ktaadn": Lost on "Some Star's Surface"

Thoreau's first trip to the Maine Woods was in 1846, while he was living at Walden Pond. Telling the story later, first in his Journal and public lectures and then in a narrative published in *Sartain's Union Magazine* in 1848, he would struggle to find a language, a form that would capture both his ecstatic interest in Maine and the sense of disappointment that he experienced there.[10] Traveling north from Bangor with his cousin George Thatcher, Thoreau hoped to secure a Native guide, but settled instead for a white lumberman, "Uncle" George McCauslin, and a young helper named Tom Fowler. Worse, where he'd dreamed of a wilderness, he found a series of rutted roads, dammed rivers, and sprawling homesteads within a largely cutover forest.[11] These places, he learns, are owned by the timber companies and a small handful of widely dispersed settlers, who want the landscape to remain not only uncultivated, but unpeopled, what land managers today call a "working forest."[12] Thoreau's challenge was to find a form with which to narrate these experiences, which utterly stymied any literary precedent.

In spite of all the logging, Thoreau was still enchanted by the landscape of the Maine Woods. On North Twin Lake, for instance, just beyond the last lumber dam, McCauslin led the party to an old campsite, navigating by moonlight. In an early draft of "Ktaadn," written in Thoreau's Journal sometime shortly after the trip, this moment stands out as a highlight of the narrative, even something of a climax. Waking after midnight, Thoreau wanders alone near the lakeshore, dreaming of moose and wolves as he looks across the water. The anticipation is palpable, even as it seems to cloud the narrator's perception, the lake becoming "the atlantic of 3 centuries ago and we the discoverers on its western shore" despite all the signs of human inhabitation and industry.[13] The days ahead would fundamentally challenge this impression, but in this moment, Thoreau is filled with a rush of wonder and anticipation. It was, in a way, the high point of the trip, an instant when his expectations and dreams seemed finally to match the scene before him, or at least his sense of it from that dark lakeshore.[14]

Of course, the real "high point" of the narrative comes a few days later, on the flanks of Ktaadn.[15] Over two days, Thoreau and his companions twice attempted to climb the mountain. On the first day, the party made a hasty camp above a steep mountain stream, and Thoreau continued alone, as though compelled upward, through the krummholz to a broad, rocky tableland near the summit:

> Having slumped, scrambled, rolled, bounced, and walked, by turns, over this scraggy country, I arrived upon a side-hill, or rather side-mountain, where rocks, gray, silent rocks, were the flocks that pastured, chewing a rocky cud at sunset. They looked at me with hard gray eyes, without a bleat or low. (*MW* 61)

Gesturing to *Paradise Lost*, mocking the insufficiency of the pastoral, Thoreau must grapple here with the register for nature, with the lack of a literary precedent for such a climb, such a place. How, he asks, can we describe this experience? How can we come to really *see* such a place and what our experiences there have meant?

The next day, Thoreau's prose is marked primarily by a sense of obscurity. Climbing again through that jumble of gnarled spruces and loose rocks, he describes being within "the skirts of the cloud," which seem to swirl and drift but never quite clear (*MW* 63). This is a place impossible to describe and impossible to even see. If the pastoral fails to provide an adequate register for this sense of nature, and the mountain remains wrapped in mystery, readers are left with only a vague sense of a harsh, though beautiful, landscape, somewhere beyond human scale, even beyond human perception. Descending this second time, Thoreau finds his companions on "the side of the peak," gathering cranberries and blueberries that grow thick between the rocks, "spicier" than at lower elevations (*MW* 66). From here, they have a commanding view to the west and the south. Even this, though, feels inadequate. If the steep ridge and the jumble of rocks seem to mock the pastoral, and the summit remains wrapped in clouds, this perspective will, ultimately, be misleading, suggesting a pristine wilderness in which it "did not look as if a solitary traveller had cut so much as a walking stick" (*MW* 66). Having just passed through this country, the illusion strikes an interesting chord. In Maine, Thoreau finds, even that clear view, so eagerly anticipated, proves misleading.

On their descent back toward the river and their batteau, the party follows the course of a stream, or, rather, a "torrent" (*MW* 67). This streambed, Thoreau realizes, like the mountaintop he's just left and the steep he climbed to get there, suggests the constant creation of the world through natural forces, the trees "barked and splintered up to their tops" and rocks "two or three feet in diameter" piled up and lodged high overhead (*MW* 67). Interestingly, even while following this obvious "path," the party is somewhat unsure of their route. In one instance, they become bewildered by an unexpected footprint, which for a moment suggests "how Robinson Crusoe felt in a similar case"—until they remember crossing this way earlier in the day (*MW* 68). All along the journey north of Bangor, Thoreau has mused on the distinction between civilization and wilderness, on his expectations and the reality of this landscape. This is the first time, however, that he begins to play with the blurring and collapse of these categories. The mountainside *seems* like a wilderness, but is it really? In the moments that follow, the answer to this question seems to depend less on the actual landscape than on one's sense of disorientation.

A moment later, the party pauses briefly to check their route:

> After leaving the torrent, being in some doubt of our course, Tom threw down his pack at the foot of the loftiest spruce tree at hand, and shinned up the bare trunk some twenty feet, and then climbed through the green tower, lost to our sight, until he held the topmost spray in his hand. (*MW* 68)

McCauslin, Thoreau adds, was once famous for such feats of wayfinding, having been a scout and spy "under General Somebody" (*MW* 69). He goes on:

> To Tom we cried, where away does the summit bear? where the burnt lands? The last he could only conjecture; he descried, however a little meadow and pond, lying probably in our course, which we concluded to steer for. (*MW* 69)

Still, even with this imperfect sense of the country ahead, the party remains somewhat unsure of their course. They are, in a word, lost.

This sense of physical disorientation is important, but it has not figured widely in critical assessments of "Ktaadn." "Disorientation" comes from the French, *désorienter*, literally to turn from the east, referring to traditions of placing and aligning churches, graves, and other sacred sites to accord with the rising sun. By thus turning away from this grounding sense of direction, "disorient" evokes both a short-term state of geographic uncertainty with pragmatic solutions (climbing a spruce tree to spot distant landmarks) and something much more unsettling. Similarly, disorient also means "to put out, disconcert, embarrass," as when one's skill with compass proves, in the end, not so helpful after all.[16] In the moment, disorientation often figures as a state of panic, less a specific fear of being misplaced than, rather, an intense sense of self-doubt and an inability to assess the situation in rational terms. Within literary prose, and particularly masculine adventure narratives, being lost often adds an element of danger where it might otherwise be lacking. Thoreau thus emphasized this dimension of disorientation in that early draft of "Ktaadn," while "discovering" Twin Lake by moonlight. In reflecting on the experience and revising the essay, however, he came to describe another kind of disorientation, an instance of curiously heightened perception in which the categories of wilderness/civilization and savage/civilized begin to blur and ultimately crumble. The famous lines that follow did not appear in the draft but rely heavily on this sense of being lost, and not in the real or metaphorical clouds, but in the thick forest below.

When Tom climbs down from the spruce, the party passes through another densely forested area before emerging in what they know to be a specific area on the mountainside, the "burnt lands." This does not, however, come to figure as a return to familiar ground, but rather one more unsettling encounter with a strange, unearthly landscape:

> We were passing over the "Burnt Lands" burnt by lightning perchance, though they showed no recent marks of fire, hardly so much as a charred stump, but looked rather like a natural pasture for the moose and deer, exceedingly wild and desolate, with occasional strips of timber crossing them, and low poplars springing up, and patches of blueberries here and there. I found myself traversing them familiarly, like some pasture run to waste or partially reclaimed by man; but when I reflected what man, what brother or sister or kinsman of our race made it and claimed it, I expected the proprietor to rise up and dispute my passage. It is difficult to conceive of a region uninhabited by man. We habitually presume his presence and influence everywhere. (*MW* 70)

With the party's wayfinding in mind, the "burnt lands" Tom was looking for from high in that spruce and the "Burnt Lands" described by Thoreau in this passage seem incongruous. Thus, this landmark proves strangely unfamiliar, more like another area to be negotiated than a true return to familiar ground. The party remains lost here, or, at least, somewhat uncertain of the direction they're headed. The "familiar" way in which Thoreau crosses this place reflects an attention to the landscape, rather than an exact sense of their route, and this new awareness, building on the sensation of being lost, suggests an earthly dynamism, the constant creation and re-creation of the world. Just as the torrent had been an agent of change, shaping the mountainside with its floods and rockslides, just as the mountain had been a "cloud factory," the Burnt Lands too are an agent of material change, a forest or a pasture emerging in response to earthly forces. Being lost, in other words, allows Thoreau to see the planet that he walks on, to better sense the various ways that it is shaped and reshaped over time. This is not wilderness, then, but a dynamic, *wild* landscape.

Turning once more to the language of myth and to Milton's *Paradise Lost,* Thoreau describes this sense of earthly dynamism, his excitement seeming to grow in each line:

> I looked with awe at the ground I trod on, to see what the Powers had made there, the form and fashion and material of their work. This was that Earth of which we have heard, made out of Chaos and Old Night. Here was no man's garden, but the unhandselled globe. It was not lawn, nor pasture, nor mead, nor woodland, nor lea, nor arable, nor waste-land. It was the fresh and natural surface of the planet Earth, as it was made forever and ever,—to be the dwelling of man, we say,—so Nature made it, and man may use it if he can. Man was not to be associated with it. (*MW* 70)

At the height of his reflections, Thoreau is still not quite sure of his exact location, still traversing a strange, unsettling landscape. He goes on:

> What is it to be admitted to a museum, to see a myriad of particular things, compared with being shown some star's surface, some hard matter in its home! I stand in awe of my body, this matter to which I am bound has become so strange to me. I fear not spirits, ghosts, of which I am one,—*that* my body might,—but I fear bodies, I tremble to meet them. What is this Titan that has possession of me? Talk of mysteries!—Think of our life in nature,—daily to be shown matter, to come into contact with it,—rocks, trees, wind on our cheeks! the *solid* earth! the *actual* world! the *common sense*! *Contact*! *Contact*! *Who* are we? *Where* are we? (*MW* 71)

These lines, referred to by many as the "*Contact*" passage, are among Thoreau's most famous prose, and have featured prominently in environmental literature anthologies, introductions to ecocriticism, and histories of wilderness and American culture.[17] This body of work has generally interpreted "Ktaadn" in terms of the Burkean sublime, in which objects of incalculable scale move the imagination to awe and a certain degree of fear.[18] Crucially, though, Thoreau's epiphany while walking awestruck through this opening in the forest, pausing now and then to pick blueberries with "a smart and spicy taste," is only distantly connected to the sense of fear that readers might expect on a cloudy mountaintop or during the steep scramble along what is today known as Abol Slide, the beginning of the Aboljacknegesic and Thoreau's presumed route down.[19] For Thoreau, who had written on Burke at Harvard, the sublime was something different entirely, in Ronald Wesley Hoag's words, "an intense encounter with an imposing physical nature that, rightly construed, was also the embodiment of an informing spirit."[20] Rather than a sense of alienation and a powerful urge to turn away or to flee, Thoreau is intensely *attracted* to this sense of an ongoing creation, seemingly more inclined toward awe and wonder than toward pleasure or fear. Christopher Hitt, writing on the potential for an "ecological sublime," suggests that this happens in "Ktaadn" when "a sublime encounter with nature seems to have the power to jolt us momentarily out of a perspective constructed by reason and language, a perspective that, in modern Western culture, has rendered nature mute."[21] That sense of a jolt or a disturbance, of coming to see the world a little differently while dramatically lost, will prove key to Thoreau's later excursions. From here, he will work to construct a new form which can accommodate these most radical insights. And crucially, this all emerges in the narrative only after Thoreau has first drafted an account of the trip. His first attempt at the essay that would become "Ktaadn," preserved in his Journal, does not contain this startling, dizzying new perspective on human experience of earthly dynamism.[22] As Hoag writes, "Rather than an immediate deflation of Thoreau's faith, the real effect of his experience atop Katahdin is retrospective, cumulative, and emphatically uplifting."[23]

Recent scholarship, influenced by the fields of ecocriticism and New Materialism, explores the role of matter in shaping and creating texts and offers a new sense of the sublime and Thoreau's attempts to capture these experiences. As Melissa Sexton puts it, "Thoreau's cries of '*Contact!*' suggest a moment when the world's materiality disrupts his preexisting distinctions and leads to questions: '*Who* are we? *where* are we?' "[24] "Ktaadn," Sexton concludes, is "an intertextual, intersocial, and materially grounded construction," itself a part of the dynamic processes of destruction and creation that Thoreau observed while on the mountain.[25] Rather than a sense of fear and failure, Thoreau glimpsed what he will later call, in "Chesuncook," "our own true *re*-creation" (*MW* 156, emphasis added). *This* is that new perspective on the world, on life at home, around which he will craft this Maine excursion. In this moment, the divide between "wilderness" and "civilization" comes under pressure, and the adventure narrative seems to crack open, the story shifting to ultimately mirror his dawning awareness of material change and "re-creation," one's self and the material world together *in motion*. This peculiar, enthralling sense of the sublime, of a wildness that continually shapes the world, emerges only after these experiences have come to their conclusion, and only when the party becomes lost in a thick spruce forest on the flanks of the mountain. Only then can Thoreau properly reflect on the experience, placing it all together in a dizzying new sense of planetary dynamism, of earthly change far beyond the human scale, accessible only fleetingly, indirectly.

Immediately after this passage, the text veers suddenly toward the prosaic. Where moments before Thoreau was lost, now he spots real landmarks: "Erelong, we recognized some rocks and other features in the landscape which we had purposely impressed on our memories, and quickening our pace, by two o'clock we reached the batteau" (*MW* 71). With this new sense of the dynamic material world in mind, the reader turns with Thoreau back toward home, and the excursion has been, from one sense, completed. Upon returning to familiar ground and turning back toward Bangor and home, the experiences on the mountain, in the Burnt Lands, and elsewhere along this journey become part of a new way of living in and understanding the world, influencing his other, contemporary projects, including *A Week on the Concord and Merrimack Rivers* and *Walden*. Having been shocked into a new way of seeing the very earth beneath his feet, these various manuscripts begin to converge around similar energies.[26] In struggling against the pastoral and the wilderness journey, Thoreau struck upon some of the themes and ideas that would shape his writing for the rest of his life. "Ktaadn" was perhaps his single most effective early essay, and it remains among his most influential works, if not his most radical. That sense of shock, of coming to see things a little bit differently, would continue to prove crucial to his emerging literary method. In his later Maine essays, Thoreau would push even harder on the purpose of travel writing, ultimately unsettling the primary categories which define the genre: home and away.

"Chesuncook": Portal to Another World

Thoreau returned to Maine in 1853, this time joining Thatcher for a late summer moose hunting expedition. He goes along, in his words, not as a hunter, but rather "as reporter or chaplain to the hunters," with a keen interest "to learn how the Indian managed to kill one" (*MW* 99). Unlike his previous excursion, this trip will require a different mode of navigating and observing the Maine Woods, a canoe rather than a batteau, and, importantly, a Penobscot guide, Joseph Aitteon, to lead the way. Almost immediately, a tension emerges between Thoreau and Thatcher, between what we might call a hunting narrative and another, less obvious kind of story, which meditates on the value of being "chaplain" on such a journey and the other ways in which one might learn to travel these waterways.

"Chesuncook" begins with the party traveling across Moosehead Lake in a small ferry, which takes them to the North East Carry. On the far end of this famous portage, they pass a group of Wabanaki hunters camped along the West Branch of the Penobscot, drying moose meat. From here, as at North Twin Lake in 1846, the narrator seems to be passing a threshold into a wilder country beyond. He describes settling into a birchbark canoe with Thatcher, Aitteon in the stern:

> Joe placed birch bark on the bottom for us to sit on, and slanted cedar splints against the cross-bars to protect our backs, while he himself sat on a cross-bar in the stern. The baggage occupied the middle or widest part of the canoe. We also paddled by turns in the bow, now sitting with our legs extended, and now rising upon our knees, but I found none of these positions endurable. (*MW* 95–6)

At first, the narrator seems to find the *process* of hunting rather striking, if uncomfortable, as it forces the party to travel in what is, for him, a novel way, to intensely attune themselves to the environment around them. The bow of the canoe offers an unanticipated new perspective on the river and that shaggy forest, a method of travel which, although difficult, suggests the wild character of the landscape. Later that day, after making camp, they will continue paddling into the twilight, and Aitteon will pull out his birchbark horn, calling for moose again and again, his practiced calls ringing out through the dark woods—"*ugh ugh ugh*, or *oo oo oo oo*, and then a prolonged *oo o o o o o o o*" in the narrator's rendering (*MW* 101). They continue like this long into the evening, calling and listening, gliding silently along, watching the darkening banks. No moose appear.

The next afternoon, the three travelers push up a tributary called Pine Stream on Thoreau's map, where they find fresh tracks along the bank. Coming to a small meadow, Thoreau hears a cracking noise in some streamside alders, and alerts Aitteon. In the lines that follow, Thoreau's prose records a series of rapid events, registering a perception that seems to eclipse the pace of rational thought. After hours of quiet attentiveness, the hunt unfolds too quickly to follow. Realizing that he sees a cow and calf moose, the narrator describes a series of impressions and connections, filling what seems, suddenly, like a crucial gap in his experience of this place, "a vacuum" in his words (*MW* 110).

The narrator hears a rustling just behind them, and Aitteon pushes the canoe rapidly backward, where they suddenly see two moose in a meadow they've just passed. He and Aitteon, recognizing their positions, duck, and Thatcher fires the first barrel of his gun over their heads. He, too, has never seen a moose before, and aims for the larger of the two. The narrator describes the rush of events after the shot, but the entire scene seems to challenge the limits of normal human perception: "one dashed across the meadow and up a high bank," disappearing and yet leaving "a distinct impression of its outlines on my mind" (*MW* 110). The other, the younger and smaller of the two, leaps into the stream and stands "cowering for a moment" and "uttering two or three trumpeting squeaks" before Thatcher fires again (*MW* 110). The calf pauses a moment and, apparently unwounded, dashes away. Thoreau recalls "an indistinct recollection of seeing the old one pause an instant" atop the bank, but he cannot be sure of what happened in these few seconds (*MW* 110). Thatcher doesn't know "whether he had fired at the same one twice or not" (*MW* 111). Indeed, only Aitteon seems equipped to comprehend everything that has just transpired, stepping onto the bank to attempt tracking them. Thoreau follows, trying to understand the signs that Aitteon points to in silence, such as "a single drop of blood on the handsome shining leaves of the Clintonia borealis" (*MW* 112). For about half an hour they track the two moose, first the mother and then the calf, finally turning back when the trail becomes thin, disappearing amid "many other moose tracks" (*MW* 112).

Returning to the canoe and continuing upstream, they find the carcass of the mother only later, half submerged in a shallow riffle. Thoreau then watches, appalled, as Aitteon skins it—"a tragical business," the "ghastly naked red carcass appearing from within its seemly robe, which was made to *hide* it" (*MW* 116). The tragedy of this scene, as it turns out, is not merely the nature of the skinning, but the gross waste of such a wondrous animal's life.[27] Thatcher keeps only the ball which killed the animal, which Aitteon recovers, flattened and deformed. Aitteon will take the hide, plus a few small pieces of meat. Only later, at Thoreau's insistence, will they return for a hind quarter, which they will leave with a group of loggers on Chesuncook Lake.

Later that evening, Thoreau sits alone in camp while Thatcher and Aitteon continue their hunting. Writing about the experience later on, this moment will form the beginning of a long reflection on the "motives which commonly carry men into the wilderness" (*MW* 119). The difference, he writes, between the "base or coarse" motives and a higher purpose is in the intention of the traveler, and in what they elect to carry home. He muses on the white pines, which, hunting notwithstanding, are the principal draw to these woods:

> Strange that so few ever come to the woods to see how the pine lives and grows and spires, lifting its evergreen arms to the light—to see its perfect success but most are content to behold it in the shape of many broad boards brought to market, and deem *that* its true success! (*MW* 121)

Hunting, like logging, is thus not merely an extractive pursuit, but obscures the real value of a journey, the lessons that one might carry home. In this moment, the "tragedy" on Pine Stream merges with logging as a combined disaster taking place across the Maine Woods. For Thoreau, the only traveler who truly knows the pine is "the poet," and, moreover, this love and appreciation for the great trees is, ultimately, their highest use—and seeing them, experiencing these forests from the bow of the canoe, comes to be the most important part of such a trip. This reflection leads to a line that James Russell Lowell, editor of the newly formed *Atlantic Monthly*, would strike from "Chesuncook" when it appeared in print some years later, in 1858: Thoreau, writing that he "sympathizes" with the "living spirit" of the pines, writes that these trees are "as immortal as I am, and perchance will go to as high a heaven, there to tower above me still" (*MW* 122). If the excursion form hinges on a certain sense of shock, of being jolted into a new awareness, "Chesuncook" suggests a new sort of relationality, a sympathy which arises with a close attunement and a gentle curiosity, or wonder. In reaching beyond the hunting narrative for another kind of story, Thoreau was also considering the potential publication of his third Maine essay, which he had drafted by 1858 and to which I will return in the following section. For "Chesuncook," however, the results were similar to Thoreau's earlier foray in magazine publishing. Lowell, striking that line, would, just as Curtis had done in the Canada essay, trace out the limits of his most unconventional thinking.[28] Although it appeared in three installments in *The Atlantic* in the summer of 1858, that pine-tree sentence wasn't restored until the text was reprinted in *The Maine Woods* in 1864.

Thatcher and Aitteon return that evening without seeing another moose, and the next day the three travelers make their way back down Pine Stream to the West Branch. From there they visit Ansell Smith's logging camp on Chesuncook Lake. They'd intended to continue to the northwest, but Aitteon learns that a party of Indigenous hunters has just killed several moose along the Caucomgomoc, where he and Thatcher had planned to hunt next. Learning this, they depart the next morning to the south, retracing their path upstream toward the North East Carry. Arriving there the next day, they encounter the same group of Wabanakis camped at the end of the portage, still drying moose meat. There is also a party of white lumberers camped nearby, whom Thatcher proposes joining. Thoreau, however, turns down this suggestion. Instead, he elects to stay with these hunters, who, he writes, "were much more agreeable, and even refined company" (*MW* 133). Robert F. Sayre refers to this evening as "Thoreau's real trophy of the trip" and the "climax" of this narrative.[29] If that early evening on the river proved an enchanting illusion, this will turn out to be the most shocking moment of the trip—the camp at the portage becoming a true portal to another way of navigating the world.

Spreading their blankets over moose hides already laid on the ground, Thoreau and Thatcher take their places beside Aitteon and a "St. Francis Indian," who later spells his name for Thoreau: Tahmunt Swasen. "While lying there listening to the Indians, I amused myself with trying to guess at their subject by their gestures," he writes, "or some proper name introduced" (*MW* 136). As in Irving's encounter with the Osages on the prairie, the narrator records a sense of surprise and delight:

> It took me by surprise, though I had found so many arrow-heads, and convinced myself that the Indian was not the invention of historians and poets. It was a purely wild and primitive American sound, as much as the barking of a *chickaree*, and I could not understand a syllable of it; but Paugus, had he been there, would have understood it. These Abenakis gossiped, laughed, and jested, in the language in which Elliot's Indian Bible is written, the language which has been spoken in New England who shall say how long? These were the sounds that issued from the wigwams of this country before Columbus was born; they have not died away; and, with remarkably few exceptions, the language of their forefathers is still copious enough for them. (*MW* 136–7)[30]

Thoreau is awestruck not merely by the personality of these people, not merely by the "purely wild and primitive" language that he is overhearing, but by the living traditions and ontologies right before him. Having read extensively in the *Jesuit Relations* and other seventeenth- and eighteenth-century accounts of Indigenous peoples, Thoreau finds himself suddenly immersed in a culture that he did not expect, even despite his readings on French Canada.[31] Amid the discourse of nineteenth-century savagism, he witnesses here a living Indigenous culture.

Aitteon and Swasen carry on deep into the night. Amid the conversation, they take turns practicing their moose calling with Aitteon's birchbark horn. After midnight, two young Penobscots join them, returning after having hunted long into the twilight.[32] The conversation continues, and Thoreau begins to ask them questions, to take notes. Spellbound, he records several place names, marveling at the specificity of the words and the differences in dialects, which arise when the four Wabanakis discuss slight variations in terminology and in place names, including that of Thoreau's Concord and other places in Massachusetts. *Musketiquid*, they correct him, is, more properly, *Musekticook*, meaning dead stream, or slow-moving river. If his reflections on the French language a few years previous had provided an alternative sense of North America and its history of exploration and colonization by Europeans, this moment proves even more arresting, even more surprising. Ultimately, though, this turns out to be merely another portal, another glimpse into a world that is not his own, but through which the experience of travel allows him to sneak a glance. To more properly see this side of the Maine Woods, to account for what has so productively, intriguingly shaken up his understanding of the world, he must return once again, and he must travel with another Penobscot guide, a seasoned traveler and teacher with the wisdom and cunning of a trickster figure.

"The Allegash and East Branch"

Thoreau's third excursion to Maine was by far the most ambitious, but it began, interestingly, by retracing the route of his second trip, departing from Bangor, crossing Moosehead Lake, and then paddling and portaging for nearly two weeks in a great loop back to Indian Island. This time, however, Thoreau is accompanied by his longtime friend Edward Hoar and a different Penobscot guide, Joseph Polis. If "Ktaadn" is something of a breakthrough, leading Thoreau to imagine new ways of telling a story which are grounded not in "wilderness" but in the material world, "Allegash" represents the most experimental of Thoreau's excursions, the foundational categories of "home" and "away" coming under pressure and ultimately reforming entirely. Indeed, "The Allegash and East Branch" was such a departure from Thoreau's previous excursions that he would decline to publish the completed text in 1858, sending James Russell Lowell, editor of *The Atlantic*, "Chesuncook" instead because he feared that Polis would read it and take offense.[33] Robert Sayre has called Polis "the most complex of Thoreau's Indian characters, indeed, the most fully developed person (after the author himself) to appear anywhere in Thoreau's writing."[34] In fact, Polis comes to be something more than a guide, more even than a central character in the narrative. He becomes, rather, a mentor who, with the cunning and wisdom of a trickster figure, orchestrates Thoreau's initiation into a new way of seeing and navigating the North American landscape and its history.

Although "Ktaadn" remains the single most widely read essay in *The Maine Woods*, recent scholarly debate has centered on Thoreau's portrayal of Native peoples, especially Polis. Sayre, whose *Thoreau and the American Indians* remains an influential study on the topic, describes in Thoreau "a development in his sensitivities which is remarkable," largely thanks to his contact with Aitteon and Polis, who, besides being guides in a more conventional sense, ensuring safety and success, "also function as the interpreters and priests who philosophically *guide* the travelers."[35] The extent of that sensitivity, however, remains under debate. Joshua Bellin, reading Thoreau's extensive notes and extracts in unpublished commonplace books, argues that Thoreau was "unable or unwilling to liberate himself from the dominant mid-century complex of beliefs concerning American Indians, their relationship to Euro-American civilization, and their ultimate fate."[36] John J. Kucich, focusing on Thoreau's experiences with Penobscots rather than his reading on Indigenous peoples more broadly, argues otherwise, writing that *The Maine Woods* reveals "an attitude that is shifting, ambivalent, and elusive, marked in turn by primitivist celebration, savagist scorn, and transformative contact."[37] Richard Schneider argues similarly that neither Thoreau's reading nor these experiences with Aitteon and Polis "ever quite disabused him of the savagist assumptions of his time."[38] Michael Stoneham, meanwhile, contests this. "Chesuncook" and "The Allegash and East Branch" demonstrate, as he puts it, an "evolving cultural critique of ethnologically established racism toward Native Americans" and an inclination "to reject the dominant savagist ideologies that were reaffirmed so often" in the texts from which he read and copied excerpts.[39]

Crucially, these various critiques converge around the fact that Thoreau's relationship to Native peoples did, indeed, evolve over the course of his life, and particularly after notable experiences, such as his travels with Polis. Kucich provides a particularly important point here, writing that Thoreau very often gets lost, "at times literally disoriented in the physical landscape but more frequently swamped in the labyrinth of an unfamiliar cross-cultural environment that took shape along this frontier between the United States and the Penobscot world."[40] In "The Allegash and East Branch," these two forms of disorientation—physical and "cross-cultural"—become one and the same because they are so carefully managed by Polis, who seems to curate Thoreau's experiences along their route, subtly "guiding" him to a new understanding of the landscape. The result is an excursion that is definitively shaped by a Penobscot worldview. Indeed, his environmental perception moves progressively closer to what Kucich identifies as a distinctively "Penobscot relationship to the environment."[41] If in "Ktaadn" we find an excursion marked by a new awareness of the material world, and in "Chesuncook" we find a portal to a Wabanaki view of New England, in "The Allegash and East Branch" Thoreau attempts to bring these two insights together. Thoreau's physical disorientation in "Allegash" and the group dynamics that emerge afterward present an entirely new means of narrating travel in remote places, in which the wilderness becomes a home ground set within a changed and still changing planet.

From the beginning, "Allegash and the East Branch" highlights the relationship between Thoreau and his guide. On the morning of their first day in the canoe, eschewing the ferry to paddle across Moosehead Lake, Thoreau tells Polis that he'd like to learn his language, and more, to really understand how he sees the world. Their agreement seems simple enough. "I told him that in this voyage I would tell him all I knew," the narrator recalls, "and he should tell me all he knew, to which he readily agreed" (*MW* 168). This compact will set the course of their journey over the coming weeks—without a set itinerary, Thoreau and Polis will discuss possible routes, and Thoreau will elect to travel through "the wildest" possible country, that he might observe Polis there, a Penobscot man navigating a remarkably rugged landscape (*MW* 232–3). Polis, for his part, recognizes his role as an instructor, and seems to see, too, where he can and cannot reach his pupil. He adjusts his pedagogy as necessary, realizing even that day that Thoreau will scorn Penobscot legends and oral history, but that woodcraft and navigation are of great interest, as well as what we might term, in his client's European-American sense, natural history.

Polis is attempting to gather knowledge, too, insights that might help him navigate the nineteenth century as a Penobscot leader. After Thoreau dismisses the ancestral story of "a mighty Indian hunter" and derides Polis as a poor storyteller, with "a drawling tone, long-windedness, and a dumb wonder," Polis turns the conversation (*MW* 172).[42] He probes Thoreau and especially Hoar, a lawyer, about a legal question, a matter of land ownership and encumbrance, specifically of rights to cut grass for hay. Underlying this question, careful readers sense that Polis is navigating a complex set of issues related to sovereignty, cultural identity, and legal claims to territorial homelands, as well as personal and community economics. He is, in a word, attempting to buy back at least some portion of his ancestral homelands. The transaction of knowledge is thus two-directional, and it is Polis, rather than Thoreau, who is most purposefully and skillfully choosing when and how to share what he knows. For his part, Thoreau is stunned to have learned that Polis is not only a respected leader of the Penobscot, having been called on by the governments in Augusta and in Washington, but also a cosmopolitan traveler who has enjoyed trips to Boston, New York, and Philadelphia (*MW* 197). Later in the narrative, when the group fractures and reforms along Webster Stream, Polis emerges as the leader of the party, and Thoreau's narration shifts to accommodate this new social structure.

Throughout this journey, Thoreau will encounter lessons from Polis, from simple observations and tricks of woodcraft to more dramatic tests of wayfinding and ontological disposition. Early in the trip, shortly after dismissing the story of Gluskabe, Thoreau is struck with wonder at a piece of phosphorescent wood, which Polis calls *Artoosoqu'*, a phenomenon which suggests vast gaps in his own knowledge of the New England woods (*MW* 180). Similarly, Thoreau asks how to navigate the woods without a compass, and Polis gives a vague answer, suggesting an attention to rocks, to the vegetation on hillsides, and to the shape and size of tree branches, bent certain directions by prevailing wind, or reaching toward the sun at a particular time of day. His answer proves unsatisfactory for Thoreau, but Polis cannot perhaps *explain* this sort of deep attention to the landscape. Instead, he tells a story of hunting with a white man, "a very good hunter," who cannot find his way back to camp. Polis takes the lead to guide this particular hunter back to camp, but Thoreau, impatient, interjects, as though not quite understanding the story: "How do you do that?" he asks. "O, I can't tell *you*," Polis replies. "Great difference between me and white man" (*MW* 185). Such wisdom, it turns out, cannot simply be explained. As the group travels onward, the narrator comes to seem more lost than ever before, not because he is in a "wilderness," for that concept has already faltered in "Ktaadn," but because he cannot yet recognize what *else* he fails to understand about this landscape.

Unlike "Ktaadn" and "Chesuncook," "Allegash and the East Branch" is organized like Thoreau's Journal, the narrative broken up by date. On Monday, July 27, 1857, three days after first launching their canoe, the narrator's misunderstanding leads to an interesting moment of tension with Polis and an afternoon spent entirely lost, pressing recklessly forward despite the terrain. The party has just passed by Pine Stream and Chesuncook Lake, the site of that "tragical" hunt with Thatcher, and they now face a famously difficult portage, Mud Pond Carry, from whence they will paddle to Chamberlain Lake. Polis will proceed over the boggy path first, carrying the canoe over his shoulders, and Thoreau and Hoar will follow with the gear. Their method, a common one among French voyageurs and still today among recreational canoeists, is to split their burden into two parcels per person, and then to "carry one half part way, and then go back for the rest" (*MW* 212). Polis disappears in the forest, but soon returns, instructing them to take a fork in the path and to follow his tracks. Thoreau suggests that he leave a bough in the path at the fork, and Polis agrees. Nonetheless, Thoreau and Hoar quickly become lost amidst a tangle of logging paths, confusing these veering log skidding trails for their portage and apparently missing Polis's sign. When Polis finds them, they are knee-deep in a cedar swamp, miles off course, jumping from tussock to tussock on a tote road that is only passable in winter. Polis is flummoxed by their confusion, but also their relentless push forward on what is obviously the wrong route.

Reunited, the three travelers discuss their options, eating a bit of bread. Polis has never used this tote road, and must return to the portage anyway, where he's left the canoe. Thoreau and Hoar, now tired and wet, are weary of redoubling their steps, and reason "that it would, perhaps, be nearer for us two now to keep on to Chamberlain Lake, omitting Mud Pond" (*MW* 217). Before parting ways, Polis casts doubt on their plan. Thoreau does not record their "consultation," but he does suggest that his guide knew better: "He supposed that the water in which we stood had flowed back from Mud Pond, which could not be far off eastward, but was unapproachable through the dense cedar swamp" (*MW* 217). A careful reader might suspect that Polis recognizes a learning moment, a chance for his clients to become more seasoned travelers.

As it turns out, Thoreau and Hoar go on to cross a ridge and enter another swamp, even more difficult to traverse. They trudge onward, finally coming to somewhat higher ground. Hoar's feet, waterlogged and blistered in his leather boots, slow him dramatically, so Thoreau pushes on alone, hoping to find Polis:

> Having gone about a mile, and got into low ground again, I heard a sound like the note of an owl, which I soon discovered to be made by the Indian, and answering him, we soon came together. He had reached the lake, after crossing Mud Pond, and running some rapids below it, and had come up about a mile and a half on our path. If he had not come back to meet us, we probably should not have found him that night, for the path branched once or twice before reaching this particular part of the lake. (*MW* 220–1)

The narrator's sense of disorientation in this moment is obvious, from the wrong fork in the portage trail to the vague sense of his position east of Mud Pond. His decision to push onward, like his seeming compulsion to climb higher and higher in "Ktaadn," reflects a remarkable degree of inexperience in such terrain. This is not, however, a productive experience in the sense of that "good reason" for an excursion, and Hoar, who catches up later, will suffer painful, tender feet for the rest of the trip. Thoreau fails to see Polis's skill in tracking his lost clients, too, both on the portage trail and later, in the swamp. He is grateful, undoubtedly, but too exhausted, too out of his element, to observe how Polis has interpreted their ill-advised march through to what he calls "Apmoojenegamook, or Chamberlain Lake" (*MW* 221). Instead, along with Hoar, Thoreau can only wash himself in the lake, spread his clothes to dry before the fire, and fall asleep on the pebbly shore, swatting at midges and mosquitoes.

This sense of disorientation, though seemingly resolved, lingers, carrying forward until two days later, on Wednesday, July 29, 1857, when the party reaches the stream that Polis calls *Madunkehunk* (Height of Land). Also known as Webster Stream, this small river has a remarkable reputation among travelers in the Maine Woods. Here again, Polis must deliberately guide his fellow travelers. Like other reaches of the Penobscot watershed, this area is heavily influenced by logging dams, as well as a canal which serves to connect the East Branch to Telos Lake, thereby linking the Penobscot and St. John's waterways, diverting flows and allowing lumberers to drive logs from this region south of Bangor, avoiding Canadian taxes and mills to the north. Webster Stream is thus an artificial waterway through a carefully managed landscape, a rapid diversion between two rivers made to serve the interests of American timber companies. Because of the increased and irregular flow, as well as the scouring effects of the logs, this otherwise small, rocky stream is particularly swift, "and can hardly be considered navigable, unless that may mean that what is launched on it is sure to be carried swiftly down it, though it may be dashed to pieces by the way" (*MW* 249). To descend from this point to the East Branch of the Penobscot, Polis must lighten the canoe and run the rapids alone, and he directs Thoreau and Hoar to an old path along the bank, where they will again carry the party's gear. Setting out, they immediately lose the trail. This is the first of three instances in which the two will become lost along this swift, rocky stream. Quickly realizing the difficulty of the terrain, by a stroke of luck they stumble back onto the path once more.

Thoreau and Hoar follow the portage for less than a mile before they hear Polis ahead of them, calling through the woods. He's reached an area of smooth water and pulled the canoe to the bank, walking about a quarter mile through the woods to find his clients. Thoreau, following him back to the canoe, observes his guide's wayfinding. Winding about between the trees, Polis can rapidly and without hesitation retrace his steps to the canoe. Intrigued, Thoreau describes his sense of puzzlement:

> This surprised me, for without a compass, or the sight or noise of the river to guide us, we could not have kept our course many minutes, and could have retraced our steps but a short distance, with a great deal of pains and very slowly, using a laborious circumspection. But it was evident that he could go back through the forest wherever he had been during the day. (*MW* 251)

For some readers, this might seem a small detail, one more example of Thoreau's interest in Penobscot woodcraft, and navigation in particular. It is, after all, one more lesson from the guide who has agreed to teach him everything he knows—and who aims to carefully show Thoreau only as much as he is ready and willing to learn. Having already lost their path once, this moment also signals the degree to which Thoreau is, in a way, already lost, even as the party returns to the canoe and travels together.

The party soon reaches another rapid, and, pulling the canoe to the bank once more, Polis disappears for a moment, scouting their route ahead. When he returns, he remarks, "You got to walk; ver [*sic*] strong water" (*MW* 253). Setting off alone in the canoe once again, he seems to intentionally leave Thoreau and Hoar to navigate the riverbanks on their own. This, we sense, is a particularly significant learning moment. The order of events in the passages ahead is difficult to follow, suggesting Thoreau's challenge in recalling, ordering, and making sense of each veering turn through that difficult landscape. It is, in Thoreau's words, "very bad walking" along the river here, he and Hoar having to scramble over great boulders, logs, and brush (*MW* 254). They quickly lose sight of each other, and Thoreau writes as though he is completely alone. This moment seems to reflect a similar experience in "Ktaadn" when, separated from his companions, he first felt that dizziness of wandering an expansive, unfamiliar, and dangerous landscape. Only here, the profound sense of solitude and smallness quickly gives way. Coming into a burnt area along the river, not unlike the Burnt Lands on the shoulders of the mountain, he suddenly encounters Polis, who has beached the canoe near a falls. Hoar is nowhere to be seen, so they climb a small ridge, and after calling again and again they finally receive a faint answer. Hoar has wandered far from the river, veering inland where he struck a faint path—possibly the portage that they'd been following earlier in the day, which might have taken them around this section of the river entirely. Hearing Thoreau's and Polis's calls, Hoar turns back to retrace his steps.

In the meantime, Thoreau, hoping, as in "Ktaadn," for a clear view, climbs a second and then a third ridge, leaving Polis behind and continuing to "halloo" for Hoar. He crosses a moose track in the forest, quite fresh, and muses on that thickly forested, rugged terrain all around him. The fire, perhaps some two years before, had cleared most of the trees, revealing three long, parallel ridges, like stony waves rising parallel to the riverbanks. Suddenly, Thoreau detects Polis in the distance, beckoning to him. Unsure, Thoreau gestures in return that he will first climb the high rock before him. The sense of isolation here—first from Hoar, and then from Polis—sets off a pattern of failed communication, misunderstanding, and poor group cohesion. From his first description of this journey through to the final draft in *The Maine Woods*, Thoreau struggles to narrate a series of veers, a world so filled with turning points and dead ends that he and Hoar simply cannot proceed forward without Polis. In this moment, wandering along an artificial waterway in a would-be wilderness, Thoreau must navigate the complexities of a rugged landscape as manipulated by humans, and his narrative must accommodate the other characters who are traveling with him, helping him find his way.

Hoar, still injured and exhausted from the long slog at Mud Pond Carry and from his detour in the woods earlier in the day, finally catches up to Thoreau. Reunited, the two reach the summit of the third ridge and look out over the burnt forest:

> We could see the lake over the woods two or three miles ahead, and that the river made an abrupt turn southward around the northwest end of the cliff on which we stood, or a little above us, so that we had cut off a bend, and that there was an important fall in it a short distance below us. (*MW* 257)

Thoreau can see the canoe as well, resting on the bank, and wonders with Hoar where Polis might be, guessing that he'd climbed a similar ridge on the opposite side of the river, also to see what lay ahead. A faint halloo ringing from the distant woods confirms this, but Polis, once they see him, seems in no hurry to return to the canoe. Remembering how he'd previously beckoned, Thoreau wonders what might be keeping him, suspecting "that there might be something more to delay him than I knew" (*MW* 257). They descend the ridge together, making toward Polis and planning to "keep him in sight" (*MW* 257). When they come near, Polis crosses the river in the canoe, and Thoreau calls to Hoar, "who had just disappeared behind large rocks at the point of the precipice, three or four rods before me, on his way down the stream, that I was going to help the Indian a moment" (*MW* 258).

The difference in perspectives here is striking, and the tension, the uncertainty that seems to grip Thoreau's prose in retelling this moment, indicates a careful, thorough reflection. Hoar, who will be separated from the party from this moment until the next morning, is simply keeping course down the stream, and has no notice of what might be happening upstream between Thoreau and Polis. Moreover, he is footsore, still suffering from that day on the Mud Pond Carry. Polis, meanwhile, is simply trying to get a hand with the canoe, waiting patiently all this time that Thoreau might help him lift their craft over the falls. Thoreau, unable to communicate with either of them, can look back on this moment with a degree of clarity, identifying exactly when he last saw Hoar and even seeing his own failure to understand Polis's repeated beckoning. On this rocky, thickly forested portage, at some remove from the path that might have led straight to Second Lake, the easy progress of a river trip—the regular succession of days, the steady pull of the current—finally collapses. The party can no longer move as one, and the way forward appears different to each of them—for Hoar, along the bank and downstream; for Thoreau, up a ridge and then around a precipice, where he might see the river ahead; for Polis, with the canoe along a very old route down along a rather infamous carry. Within the text, the question becomes how to narrate this general confusion, how to describe the breakdown of communication, the experience of disorientation. This, ultimately, is the challenge of the excursion form. If the "good reason" of this trip is *this* moment and the lessons that begin to crystallize only later, the sense of confusion becomes key to the new, somehow larger sense of the world that emerges afterward. Thoreau's challenge is to recreate this experience of disorientation and uncertainty, to bring the reader through the same experiences that he is struggling to reconstruct.

Reunited with Polis, Thoreau lends a hand, writing that he "helped get the canoe over a fall, lying with my breast over a rock, and holding one end while he received it below" (*MW* 258). After some ten or fifteen minutes, Polis launches the canoe again and sets off alone down the river. Thoreau hastens to catch up with Hoar but quickly realizes that he is nowhere to be found. The panic that ensues registers a sense of shock, a jolt that will eventually bring Thoreau to a new sense of the landscape. It was as if Hoar had "sunk into the earth," or,

Thoreau fears, fallen and injured himself among the rocks or tumbled into the river (*MW* 258). Polis, reaching another waterfall, is ready to make camp, but Thoreau is now gripped with panic. In the version of the narrative which was published in 1864 in *The Maine Woods*, just as in the more raw, immediate drafts in the Journal, Thoreau imagines "only the worst" (*MW* 259). Shouting and searching in the fading light, he can think only of "what I should do the next day, what I *could* do in such a wilderness, and how his relatives would feel, if I should return without him" (*MW* 259).[43]

This moment is a turning point in the excursion, the definitive instant in which Thoreau's reckless disregard for the terrain and for Indigenous relationships to the land comes most startlingly into view. The solitary climb above camp in "Ktaadn" and his mad dash into the clouds the next day, the stubborn insistence on continuing along the tote road north of the Mud Pond Carry, these and other moments demonstrate what Polis seems to regard as a strange, dangerous, and harmful relationship with the landscape. Realizing that Hoar is lost (again), Polis steps in as a guide, but his reassurance does little to calm Thoreau, who has proposed firing a gun, lighting a signal fire, and continuing desperately down the river. Polis has seen Hoar's tracks, and calmly refutes each of these ideas, noting that Hoar, who has just returned from California and has significant experience in remote landscapes, can simply make his own camp. They will find him in the morning. Thoreau, still anxious, agrees, but his anxiety lingers through the night, rising and falling as he imagines Hoar's voice over the rapids or pictures him hurt among the rocks.

The next morning, Thoreau demands an early start. Polis, still confident that Hoar is safe, plays along. And sure enough, they find Hoar shortly downstream, smoking his pipe in a campsite along the East Branch, having passed a comfortable night alone. When Thoreau sees him, he is overjoyed, calling again and again. Polis seems practically to roll his eyes, remarking, " 'He hears you,' as if once was enough" (*MW* 262). From this point on, the three travelers make their way smoothly downstream, with no more incidents at the portages or elsewhere. Group dynamics immediately change, too, and Thoreau acknowledges a new affect among the group, a sense of cohesion that wasn't present before. In his draft of the essay in the Journal, he adds a line: "Perhaps the beauty of the scene was somewhat enhanced to our eyes by the fact that we had just come together again after a night of some anxiety."[44]

Despite this new feeling, friction does still exist between Thoreau and Polis, particularly within the earliest draft of the text. When Polis kills a cow moose, for instance, Thoreau is openly critical, harkening back to his reflection on hunting in "Chesuncook." Polis wants to remain in the area to wait for the calf, who he expects to see soon, but Thoreau expresses impatience. In the Journal draft, he admits "I threw cold water on his hunting."[45] Still, even this spat does not entirely seem to ruin this new feeling that emerges within the group. Thoreau's revisions to this passage, at least, indicate a desire to smooth over what was obviously a tense moment: "I had a short ~~argument~~ talk with him about killing moose for their hides—in which he used the common white man's argument about the necessity of supporting his family."[46] Polis, patient but unapologetic, explains that skillfully skinning and preparing a hide can fetch him a significant amount of money, allowing him and some other Penobscots to buy back ancestral lands, a point that he has already shared.

Leaving Webster Stream behind, new insights begin to emerge, and Polis's various lessons seem to settle with Thoreau, and perhaps Hoar as well. Having seen the dangers of this landscape for themselves, they are now prepared to navigate this watershed with Polis in a new way, and thus to learn from and with him. As Sayre notes, Thoreau "repeats the story of Polis's nearly starving" as a child, when he and two Native companions were suddenly overtaken by an early snowstorm and freezing conditions, forcing them to make a long, challenging march back to Oldtown.[47] Polis was eventually saved by "a white teamster with supplies, who, seeing their condition, gave them as much of his load as they could eat" (*MW* 279). The sudden interest in this ordeal reflects a new respect for one's reliance on others, and also for the terrain itself. It also reflects a dawning awareness of this wild landscape as a Penobscot homeland, a place where the traveler might find a campsite nearly anywhere along the river, and where stories from Penobscots and others can crucially inform one's navigation. Regardless of the degree of Thoreau's sympathy with Indigenous communities, we might see in such moments, at the very least, an impulse toward some new way of seeing things and, moreover, a testament to the difficulty of shaking off one's prejudices.

Interestingly, this moment has not featured prominently in criticism on "The Allegash and East Branch." Sayre calls the whole episode "an initiation," almost too grand to have been orchestrated by Polis, and perhaps unrecognized even by Thoreau, but nonetheless sudden and "decisive."[48] This moment is, indeed, a threshold, and having crossed it, Maine begins to look quite different, becoming a storied landscape that exists outside the bounds of settler time, and less like that notion of "wilderness" that Thoreau seeks in "Ktaadn." Kucich concludes that "both Polis and Aitteon initiate Thoreau into contemporary Penobscot politics only at the end of their journeys, after establishing a relationship and, more importantly, after carefully guiding him through the extensive territory that the Penobscot still consider their own."[49] In "Chesuncook," Thoreau ends with a visit to John Neptune, the governor of Indian Island, and in "The Allegash and East Branch," paddling into Oldtown together, Polis gives a detailed account of the Old and New Parties, which are deeply divided along the lines of religion and education. In both of these essays, Kucich concludes, Thoreau "leaves on a primitivist excursion to the deep woods and returns to an Indian Island that is at once fully rooted in Penobscot tradition and very much a part of modern America."[50] The extent to which Thoreau recognizes and appreciates this change hinges, above all else, on his finding his bearings after, in each trip, becoming lost.

The excursion form, in *The Maine Woods* and in other texts examined here, is structured around a journey out and back, from "home" to somewhere "out there," where a narrator encounters some startling, unsettling, disturbing new insight, perhaps a new sense of the world entirely. Only back at home, Thoreau already knows, can the most significant lessons take their final form. Polis, however, leads him to see things differently, even here. On the main branch of the Penobscot, just above the primary log boom where the great pines are funneled off to the mills, Thoreau turns to his guide and asks, "if he was not glad to get home again." Polis's reply subtly reflects a startling reversal of these deeply grounding categories of Thoreau's life, travels, and writing: "It makes no difference to me where I am" (*MW* 296). For Polis, "home" is just as much "out there" as in his white house with the blinds, tidy gardens, and fruit trees. He belongs equally to both places, and is as comfortable in either, not as a matter of adventurous disposition or curious nature, but as a matter of ontology. Likewise, he is just as "at home" in Boston, New York, or Philadelphia as in these ancestral lands. Because of this, he is capable of showing Thoreau, who seems from this perspective a rather slow pupil, an entirely new way to navigate the world, not merely by following footsteps back toward "home," but by reimagining the shape of his journey in time and in space and beginning to more precisely see whose lands he is traveling through.

Ultimately, *The Maine Woods* represents an evolution in Thoreau's use of the excursion form. In "Ktaadn" he revises a somewhat disappointing adventure narrative to accommodate a stunning realization of earthly dynamism, of his positionality in relation to an ongoing creation and re-creation of the world itself. In "Chesuncook," a wonderful means of navigating the rivers and lakes of Maine, floating in a canoe by starlight, intensely aware of the dim forests all around, comes to a ghastly conclusion, but is saved by another surprise, a glimpse into a living Wabanaki culture that leaves him utterly spellbound. This third narrative, messy and complicated as it proves to be, opens his eyes yet further, building on those previous journeys to suggest a world that is, as he wrote to Blake shortly afterward, "a little larger, and not, as usual, smaller and shallower."[51] More than anyone else, Polis—as the central character of "The Allegash and East Branch"—shows us the most radical, unsettling potential of the excursion form. In this startling reversal, Thoreau finally comes all the way back toward home. "The Allegash and East Branch" is thus Thoreau's most radical travel writing, anticipating the emergence of Anthropocene nonfiction and the pressing need to understand the planet as an interconnected human-natural system, a changed and changing landscape that we cannot navigate without a good guide and clearheaded companions. In his experiments with disorientation and wayfinding, Thoreau thus shows readers how to write and think about our own experiences in the twenty-first century, offering not a blueprint, but a method for finding our way in a confounding and still marvelous world.

Coda: Excursions in the Anthropocene

In *Horizon* (2019), Barry Lopez writes of a storm approaching the Oregon coast. He is camped in an old log landing, an area within a clearcut where old-growth trees were once stacked in great rows before being trucked away. The landing is an area of compressed soil and relatively bare earth, surrounded now by a riot of new growth, young trees replacing those that were hauled away, part of a new, quite different ecological community beginning to take shape. Lopez has camped here before. From this spot, some 150 miles from home and 600 feet above sea level, half a mile inland, he has an unobstructed view of the Pacific, a perspective that has drawn him again and again to these logged-over woods. The log landing is a place to think, but not in a conventional sense. Rather, in this setting he finds himself suspended in what Inupiaq Eskimos call, in translation, "earth and the great weather," and what I will refer to here as simply the planet.[1] Amidst all of this earthly unfolding—the regular storms, the clearcut forest and its steady regrowth, the comings and goings of animals, the history of this particular coastline for Indigenous peoples and for explorers like James Cook and Meriwether Lewis—Lopez identifies "the right setting" "for getting out of myself and trying to address the questions I had."[2]

This book has argued that excursions reflect a disturbance to one's more settled way of being in the world. In these texts, authors experiment with techniques such as digressions, genre shifts, and the formal and affective expectations of readers. Through each of these and other literary techniques, texts attempt to make sense of a world that seemed to be in the midst of dramatic social and ecological change. More than a century later, Barry Lopez wrote similarly, his books unsettling readers' expectations about places and cultures around the world. Best known for *Arctic Dreams: Imagination and Desire in a Northern Landscape* (1986), which won the National Book Award in Nonfiction, Lopez traveled widely, often collaborating with scientists and artists and others to better see and understand places around the world. Written at the end of a career spanning more than fifty years, *Horizon* takes all this to a new level. Rather than recreate a single journey or some series of encounters in one geographic place, *Horizon* looks back at a lifetime of travel to explore the pressing questions that remain, visiting Oregon, the Galápagos, equatorial Africa, Australia and New Zealand, and finally Antarctica, with many meanders along the way. To read this series of journeys and reflections, to muse with Lopez on the meaning of his travels, is to face an existential question that emerges already formed in the book's prologue, without any explicit gesture toward climate change, environmental collapse, or human extinction. What, Lopez asks, will happen to us?

Horizon is best understood in terms of the excursion, though its structure and the scope of its travels suggest a particularly extreme swerve, a dramatic series of veers out and back rather than a single trip. Like other texts examined in this book, *Horizon* is an experiment, a work of autobiographical travel writing that asks us to revisit our expectations. Like Fuller's *Summer on the Lakes*, Copway's *Life, History, and Travels*, and Thoreau's *The Maine Woods*, the book is retrospective. Moreover, *Horizon* also insists on the significance of lived experience, but the book does more than simply reaffirm the value of travel in shaking us awake. Unlike those earlier books, *Horizon* emerges over a period of many years at the end of Lopez's life, when he revisited his travel notes and other writings, returning to consider the questions that remain, the ideas that continue to disturb him long afterward.

Throughout *Horizon*, Lopez insists on the specificity of particular settings, on the intricacy and complexity of what scholars have come to call place. Earlier writers anticipate this, though none have the resources, technical or otherwise, for taking it to such extremes. The sites within *Horizon* become resources for thinking, but they are also themselves open to interpretation, open to being known in some more intimate and elusive sense. And they are immensely difficult, perhaps impossible, to fully comprehend. To read *Horizon* is to think with these places as much as with the narrator, to consider the ways that travel, that place itself, can produce meaning. In the structure of the book and in the physical objects within it, which come to operate as keys and signposts, Lopez suggests that we reconsider the act of reading itself. Within *Horizon*, text and material world come to operate as coproducers of meaning.

In addition to charting a route across much of the world, *Horizon* also looks back on a life of writing, on the ways in which meaning comes to emerge from experience and from specific places. Two important points emerge from these reflections: first, Lopez notes a general working method of going away to see something and then to report back, as a messenger or a "courier" with a certain set of obligations; second, Lopez realizes, even early on, that readers' sense of his writing might not match his own, and that he must therefore trust us, to some degree, to veer home again with the most important conclusions. In the introduction, "Looking for a Ship," Lopez reflects on his childhood in New York and California, narrating his early interest in travel and the choices that led him to a life of writing, of attempting to capture the kinds of insights that he gleaned in distant places, first through photography and then in literary prose. The point of the book, though, is not to narrate a life or to chronicle specific visits, but rather to create a space in which the wisdom of travel, the meaning of experience, can unfold within a literary narrative. The text is thus associative rather than merely narrative, reflective rather than chronological. *Horizon* is rife with gaps and temporal jumps, the text shifting from present-tense narration to a recalled memory of another journey somewhere far away and then speculating on a human future. The purpose of this, of illuminating these connections in order to address some key questions, becomes a kind of key for readers.

Lopez describes his early career in terms of "incessant" traveling all over the world, of seeking vivid experiences in distant places as an alternative and supplement to continued university education. As he gained experience, however, his method came to reflect a sophisticated understanding of the writer as an intermediary between the reader and the material world. He describes feeling "an ethical obligation as a writer, in addition to an aesthetic one."[3] His purpose, he came to believe, was to travel widely and to report home, to present his experiences and his prose as a gift to the reader. He came to see, too, that an author must not be prescriptive or presumptive in their conclusions, must accept, in other words, that the readers' conclusions might transcend their own. In this sense travel and literature, life and art, seem to merge in an effort to better understand ourselves and our world. He goes on:

> I saw myself then, as a sort of courier, a kind of runner, come home from another land after some exchange with it and its denizens, carrying, by way of a story, some incomplete bit of news about how different, how marvelous and incomprehensible, really, life was, out beyond the pale of the village in which I had grown up.[4]

In retrospect, he admits, this way of working seems dangerous—if not delusional then at least overly self-important, leaving one balanced on the edge of an overwhelming commitment to their work. On the other hand, he wonders, how can one *not* take the world with an almost deadly degree of seriousness?

This method, however overly grave it might have seemed even to Lopez himself, suggests an important point for contemporary readers. The pace and scope of Lopez's travels, though perhaps enviable to some, are not necessary for the kind of writing and meaning-making that takes place within *Horizon*. The point, rather, is that his writing provides some other means of seeing and understanding the world, some resource for living, inviting us to consider meaning-making as part of a broader ecology that exists within the text and outside of it. In *Horizon*, experiences, texts, and Lopez's return visits alike contribute to broader systems of meaning in which the landscape merges with the text, creating something still wider, broader than the sum of its parts. Travel, Lopez thus suggests, has the capacity to bring together the literary and the material world across an expanse of time, to combine elements of these two realms of meaning-making together in a process that is far larger and more complex than the book in our hand might first appear. To consider the meaning of travel, in other words, is to engage in a literary act, a process of reflection, arrangement, and revision in which we come to understand not only something about ourselves, but about the world and our place in it.

Horizon is structured around a series of six extended essays, each of which is in turn built around a set of experiences in a particular geographic place. The first, "Cape Foulweather," describes that log landing on the Oregon coast. In the winding reflections that follow, a quiet, wet forest becomes a site of invention and connection, of profound thinking that stretches far beyond the spatiotemporal bounds of an evening or a day or even a series of days in an aggressively logged-over forest. The networks that unfold from here—stretching backward to the sixteenth century and even earlier and forward to an uncertain future, moving geographically out and back, delving deep into the narrator's own memories and following whatever paths we might find there, into whatever intellectual territory they might lead—return, again and again, to the campsite, to that sweeping view of the Pacific. The log landing where Lopez has pitched his tent almost becomes a stand-in for "home," even as it never quite loses its sense of mystery. Moreover, his recalled journeys from other points in his life all come to operate within the text as a series of excursions in their own right, each one looping out and back again within the bounds of the essay and then returning to that camp, the storm always on the horizon. By learning to navigate this text in terms of these micro excursions, by finding our way within a series of looping, disturbing returns to this present moment on the Oregon coast, readers come to see a method of reading and writing that is uniquely fit for the uncertainty, the darkness of what Lopez establishes as our present moment. *Horizon* thus pushes the excursion to its most developed, powerful form. By opening with that obtrusive, unshakeable question—"What will happen to us?"—and by establishing a structure which depends upon and yet also departs from physical journeys in the material world, the book challenges readers to find our way, to establish our own meaning when the future seems uncertain. These departures and returns are key to the structural integrity of "Cape Foulweather," and to the book itself.

As a literary form, the excursion is structured around a narrative swerve, the text veering away from and then in some way returning to the narrator's home with new insight on what is out there, just beyond the everyday view. This movement out and back brings the unsettling or disturbing elements of travel to the foreground. The point of an excursion, and what ultimately separates these texts from other forms of travel writing, historical or contemporary, isn't to recreate oneself or to reaffirm some already established means of navigating everyday life, but rather to bring that new view, that new sense of the world, back home again. Lopez, whose literary career spanned seven continents over more than five decades, made this central to all his writing. *Horizon*, however, stands out for its sense of urgency, for the immense scale of both its wanderings and its lessons. By highlighting the value of these disturbances for contemporary readers, Lopez demonstrates that travel, and literary writing about one's lived experience, can continue to inform us in the twenty-first century, can continue to offer new insights about the dynamic worlds unfolding in the Anthropocene.

In the nineteenth century, when authors like Fuller and Copway and Thoreau were traveling and writing about their journeys in literary prose, the tourism industry was only just emerging in the northeastern United States. Already, though, the experiences of certain places, such as Niagara Falls and the Adirondacks, had become scripted, first by prominent artistic depictions of the scenes and then by various amenities for visiting travelers. In the decades that followed, organized tours and guidebooks and later increasingly strict official regulations further standardized the experience of these and other destinations throughout the country. This trend continued in the twentieth century, as automobiles and the American highway system allowed millions of travelers from around the world to experience places that were officially "protected" and curated as national parks and other official sites. Thus the one-time US National Park Service employee and writer Edward Abbey famously critiqued "industrial tourism" within the parks, advocating a less congested, more authentic experience by way of government policy and regulation—preserve the experience of the parks, he argued, by outlawing motorized travel, requiring visitors to access these areas on foot or, in some cases, by horseback or bicycle.[5] This critique, and others before and since, drew from and contributed to an influential movement for "wilderness" preservation, as well as a large subculture of backcountry travelers who hiked, paddled, rode horses, and skied in remote areas as a means of avoiding crowds and experiencing a more vivid connection to the landscape. This trend continues today, but tourism within iconic parks such as Yellowstone and Yosemite continues to grow, creating complex logistical and management problems. Still, even the most dedicated backcountry travelers admit that one "must" see certain sites. In the twenty-first century, these places and our journeys to see them, particularly in the form of the road trip, have become iconic symbols of the United States and of American environmentalism, even amidst growing pressure to reduce the environmental footprint of individual travel and of the tourism industry.

Even as the experience of these and other places around the world becomes streamlined to merely a checkbox, one more tick on a list of "must-see" sites, American writers and readers continue to believe in the transformational power of travel and of literary writing about these journeys. For many, the allure of travel, of immersion in a completely unexpected world which one is not quite equipped to comprehend, remains strong today. This is why study abroad programs remain popular among high school and college students, and why trends like #vanlife, #digitalnomad, and #fulltimetravel appeal to so many, and to young people in particular. In a culture where many feel that authenticity and genuine connection to the material world are under threat, experiences outside of one's everyday life offer a renewed sense of wonder at the larger, ultimately incomprehensible world. Travel, in this sense, is not merely a means of escape or retreat from the mounting pressures of one's "normal" life, or even a means of redefining "success" or identity. Rather, travel represents what Lopez identifies as an alternative and a complement to the sort of learning that takes place in universities, where "domestic comforts and unconscious detachment from the workaday world" turn the classroom into "an unsafe place in which to remain … in spite of the intense stimulation that far-ranging, learned conversations there always seemed to provide."[6] Reading *Horizon*, one comes to sense how the nature of *this* sort of travel not only stimulates and informs, but sticks with the traveler long after, lingering with them far beyond the end of a semester abroad or a year's wanderings. This sense of awakening, of coming to see the world from somewhere beyond or to the side of one's limited cultural sphere, is more important now than ever before, and we needn't travel around the world to experience such moments. Rather, we might cultivate an openness to the ways in which meaning can unfold in diverse, unexpected ways, perhaps in our everyday environments or in art. To do so, we might begin by attending to the surprises that give rise to some new sense of the world.

In "Cape Foulweather," Lopez describes waking in his tent to gentle, nonhuman sounds. He describes what feels like a dream, an unearthly experience in which this landscape becomes more mysterious than ever before: "I listened hard for clear definition, but the disturbance was difficult to characterize, even when I came fully awake. These were sensations occurring between a sound and a feeling, but unmistakably there."[7] When he unzips the tent door a moment later, he sees five Roosevelt elk in the moonlight. These animals, looking back at Lopez in the dark, are a striking, if not exactly unexpected encounter. For Lopez, though, they become guides to some new means of seeing the world around him. Rather than returning to the tent, he steps out to quietly, respectfully follow them, feeling his way along an old gravel road by starlight. In that darkness, his tent fades until it seems "like some sort of space vehicle, to which I was attached by a tether." In this brief, quiet moment, Lopez becomes an alien figure in a world peopled by "these mute elk," and the silence comes to suggest "something imminent," as though the reader, along with Lopez, has reached a "crossroads between my own and another world."[8] These elk, the silent darkness, the undeniable, ultimately inseverable connection between Lopez and his tent and the truck all suggest another world that we as (modern, Western) humans cannot quite access. Lopez reminds us in this passage that there is always a radically different perspective on the environment around us, and that occasionally, we can catch glimpses of these infinite other ways of seeing the world. We simply need to be attentive to the "disturbances" that are always there, somewhere "between a sound and a feeling." Moreover, we must carry these disturbances with us, that their meaning can continue to unfold.

Later in "Cape Foulweather," Lopez sits in the log landing in bright sunshine, reading a scientific article about the nature of light. While reading and thinking, a spruce forest catches his eye, a forest he has known, but which seems unnavigable, "a thing apart."[9] There is a dirt track, an old skid road, along its perimeter, and an old clearcut to the south, now recovering. Beneath the spruces, however, there is almost no sunlight. As though the article in his hands is suddenly insufficient,

Lopez departs from the log landing, making his way toward this "thing apart." The elk and that other world, just barely visible in the dark of the night, simmer in the background of this moment, as we approach the dark forest.

Just fifty feet into the wood, Lopez can barely see the clearcut behind him, where bright sunlight glares. He is traveling, he reflects, like a wild animal, moving just inland from the clearing, where "an exposure might compromise them."[10] Fearful of losing the backlit trees behind, of becoming lost in those thick, dark spruces, he nonetheless continues. Anxiety increasing, he describes reaching an "emotional limit" as the darkness settles in, and yet he persists, sitting down at the base of a tree, facing back the way he has come. He describes a practice here that seems almost a kind of meditation, an act of sitting with one's fear, embracing it as he feels out this darkness and his reactions to it. A sense of calmness gradually comes over him, and the woods become at once "Earthly" and evocative of "the darkness of the expanded universe, the unbounded Empty Nothing that the Universe was hurling itself into at the speed of light."[11] In this moment, Lopez's experience of the dense spruce forest is opening into something else. The log landing is becoming a point of departure in another sense. From the immediacy of experience, the text will move from what might at first seem a nonhuman experience of the wood, a travel corridor for those elk and other animals, to a consideration of the inhuman universe, and from there back to humanity itself, that existential question nearly coming to the surface again. Lopez struggles to hear any sound, to detect any movement in the dark, and then, suddenly, the text seems to explode outward, the darkness ahead of him becoming something much more pressing, more threatening. He is suddenly overwhelmed by anguish, the same "agony and despair that I perused nearly every day in international reports about the fate of nameless thousands of the defeated."[12]

This moment illustrates the dramatic veers that take place regularly throughout *Horizon*, in which Lopez seems to defy the neat boundaries between the humanitarian and the environmental, between discrete places and these more global, more planetary themes. When he leaves the tree where he has been sitting, his "watchtower" over "a space seemingly infinite in every direction," he moves as though emerging from a meditative state, describing a slow process by which he "reentered a world of differentiated objects, conscious again of the force of gravity."[13] He is, strikingly, no longer afraid of the darkness, feeling "weirdly cleansed of the ancient cowardice that causes us—for reasons of self-preservation, they say—to turn away from the suffering of strangers."[14] Moreover, he is unable even to condemn anyone for this suffering, to feel indignation. Instead, having thus encountered this darkness, he confesses a readiness to return home. In one sense, he is ready for a more comfortable state of existence, away from the dark spruces and the threatening encounter there with this overpowering darkness. On the other hand, this turn toward home suggests a sense of reorientation, as though emerging from the woods allows Lopez, and the reader as well, to more fully fathom the clearcut, the darkness beneath the spruces, and human suffering around the globe.

When the darkness amidst the spruces is transformed, Lopez broaches a powerful point about experiencing the world in the twenty-first century. This space in the wood remains a site of deep, even threatening mystery, but the "darkness" is now distinctively human and decidedly global in scale, always impossible to fully observe and yet always inescapable. In the Anthropocene, he suggests, to travel is to encounter sites of tremendous horror, and to see the ways in which new worlds are already emerging, like the new ecosystems taking shape following a clearcut. The question here is about the place of humans in this emergent world, or, perhaps more poignantly, about how a single person can navigate this moment. This passage still rings with that question—"What will happen to us?"—but here the answer threatens to leave us entirely unmoored, because in this moment, "darkness" becomes a manifestation not simply of human suffering or even of human evil, but also of the "indifference" of those who turn away. This is not, however, the only form that humanity will take within *Horizon*, nor are we left with the despair that we might expect when broaching such questions. Confronting this darkness leaves us both implicated and impotent, but it also manifests an act of profound faith, a commitment to some other, not yet apparent way of being in the world. Writing and thinking about the disturbances that shape our day-to-day lives, Lopez suggests here, can help us foster new ways of living and being in the world.

Travel remains life-changing, for Lopez as for others, insofar as it is disturbing, challenging our established ways of seeing and navigating the world. *Horizon* meditates on these disturbances, and it challenges readers to not only keep up with a complex series of arrivals and departures, to address the questions that have lingered, but also to reread our own experiences through a similar method of reflection. Early in *Horizon*, discussing his method in composing the book, Lopez gestures directly to readers, and perhaps to younger readers in particular, on the verge of their own journeys:

> My further desire in planning this book was to create a narrative that would engage a reader intent on discovering a trajectory in his or her own life, a coherent and meaningful story, at a time in our cultural and biological history when it has become an attractive option to lose faith in the meaning of our lives. At a time when many see little more on the horizon but the suggestion of a dark future.[15]

There is a certain lack of originality, he admits, in looking back over one's life. Nonetheless, there remains a radical potential in thus creating meaning, in drawing from the affordances of the literary to invest in one's life, one's world, a sense of greater purpose and direction. In the present, Lopez argues here, we need these coherent and meaningful stories more than ever. When an image or a piece of music or the book in our hands brings us together in some way, we experience an ecology of meaning, the specific ways in which art brings us outside of ourselves, (re)connecting us to the wider world. *Horizon* takes readers around the world, but ultimately, it leads us out further yet, out beyond the page. To read this book is to be invited to consider interactions between text, culture, and material world.

Michiko Kakutani, reviewing *Arctic Dreams*, describes an "impressionistic, collage-like picture of the North" in which "narrative digressions gradually take on a kind of organic order, as the author loops back and forth from the philosophical to the scientific, the metaphoric to the specific."[16] These images of specific places become key to the structure of *Arctic Dreams*, these regular, looping departures and returns lending the book its final shape. As a result, *Arctic Dreams* resembles some works of fiction in that "it treats the distant, snowy world of the Arctic as a place that exists not only in the mathematics of geography, but also in the terra incognita of our imaginations."[17] *Arctic Dreams* is an excursion, and it moves homeward with an unshakeable sense that the future Arctic is threatened by a complex of wicked problems, from colonialism and resource extraction to globalism and modernization. To read this text is to see another example of what it means to depart and to return home in the late twentieth century, with a story about the marvelous, ultimately incomprehensible world out there, just beyond our everyday horizon. *Horizon*, however, aims for something different.

Returning to the Arctic in *Horizon,* readers again witness a world that faces an existential threat. Here, however, the stakes are greater, for the world is different in the twenty-first century, and Lopez's questions have accrued and matured with the onset of what he calls elsewhere "an era of emergencies."[18] *Arctic Dreams* probes the connections between landscape and human imagination, considering the vastness of the Arctic by attending to panorama and suddenly then to minute detail, to our long obsession in the West with this region beyond compare. Today, of course, to observe the Arctic is to confront a wounded planet, to wrestle with climate breakdown and a series of cascading disasters. Any place becomes, ultimately, a window on the world. In *Horizon*, the deep, obscure nature of places around the world become, in the end, a window on the human imagination, on our capacity for wonder and creativity and compassion, but also for cruelty and violence. Likewise, we confront a darkness, a sense of uncertainty, that is more pressing than ever before, more threatening, and for which our usual methods of navigation might not prove useful. What, in the end, will happen to us?

Against all of this, the landscape resonates, suggesting new ways of being in and navigating the world. To access this, to understand how we might reimagine *who* and *where* we are, is ultimately an act of faith and of storytelling, of experimenting with the kinds of stories and the kinds of worlds that we want to build. The log landing at Cape Foulweather, set amidst a scarred, second-growth forest, with that unobstructed view of a coming storm, is at once a symbol of the broader world and our present moment *and* a specific place from which to look out, both a destination and, in the end, a "point of departure."[19] *Horizon* suggests that the acts of reading and writing are not merely an exercise for the dangerous comforts of a university classroom, but a starting point for life in this new world. Thus, when we sit with Lopez amidst the clearcut, in a landscape that seems like a new "creation, unknown and unplanned," we come to see that there is no stark divide between the log landing and the words on the page. In *Horizon*, rather, the real value of travel emerges only in the interplay between text and world, between human imagination and our home ground, the planet itself.

Notes

Introduction: The Excursion

1 Henry David Thoreau, *Letters to a Spiritual Seeker*, ed. Bradley P. Dean (New York: Norton, 2005), 153, 153–4. Thoreau, like most other nineteenth-century US writers, used "Indian" as a blanket term for all Native nations and peoples of North and South America. In this book, I refer to specific tribal affiliations whenever possible to reflect the ongoing struggle for Native cultural persistence and sovereignty. I also refer more broadly to Indigenous and Native peoples and cultures in general, in contrast to settlers.

2 The Anthropocene remains contested across a variety of fields, including in the humanities. For example, starting points for this proposed era range from the development of nuclear technology to as far back as the dawn of agriculture in the Neolithic era. On the critical implications of these various starting points for scholars in the humanities, see J. R. McNeil, "Introductory Remarks: The Anthropocene and the Eighteenth Century," *Eighteenth-Century Studies* 49, no. 2 (2016), 117–28. See also Dipesh Chakrabarty, "The Climate of History: Four Theses," *Critical Inquiry* 35 (2009), 197–222. For an influential critique of the Anthropocene grounded in studies of colonialism and slavery, see Katherine Yusoff, *A Billion Black Anthropocenes or None* (Minneapolis: University of Minnesota Press, 2018).

3 Lawrence Buell, *The Environmental Imagination: Thoreau, Nature Writing, and the Formation of American Culture* (Cambridge, MA: Harvard University Press, 1995), 115. For a recent defense of Thoreau and a summary of one notable critique, see Jason Mark, "In Defense of Henry David Thoreau: New Literary Revisionism Might Make the Seer of Walden Pond More Relevant to Readers Today," *Sierra*, October 27, 2015, www.sierraclub.org/sierra/2015-5-september-october/green-life/defense-henry-david-thoreau.

4 Buell, *Environmental Imagination*, 116.

5 Laura Dassow Walls, *Seeing New Worlds: Henry David Thoreau and Nineteenth-Century Natural Science* (Madison: University of Wisconsin Press, 1995), 5.

6 See Lance Newman, *The Literary Heritage of the Environmental Justice Movement: Landscapes of Revolution in Transatlantic Romanticism* (New York: Palgrave Macmillan, 2019), 194–212; Cristin Ellis, *Antebellum Posthuman: Race and Materiality in the Mid-Nineteenth Century*

(New York: Fordham University Press, 2018), 61–95; and Rebecca Kneale Gould, "Thoreau, Race and Environmental Justice: Deepening the Conversation," *Concord Saunterer* 25, (2017), 171–83.

7 See Kristen Case, "Knowing as Neighboring: Approaching Thoreau's Kalendar," *J19: The Journal of Nineteenth-Century Americanists* 2, no. 1 (2014), 107–29; and Sarah Dimick, "Disordered Environmental Time: Phenology, Climate Change, and Seasonal Form in the Work of Henry David Thoreau and Aldo Leopold," *ISLE: Interdisciplinary Studies in Literature and Environment* 25, no. 4 (2018), 700–21. See also Kristen Case, *Henry David Thoreau's Kalendar: Charts and Observations of Natural Phenomena* (Minneapolis, MN: Milkweed Editions, 2025).

8 See James Finley, "Matter and Objects in *The Maine Woods*," *Concord Saunterer* 22 (2014), 112–19; Rochelle L. Johnson, "'This Enchantment Is No Delusion': Henry David Thoreau, the New Materialisms, and Ineffable Materiality," *ISLE: Interdisciplinary Studies in Literature and Environment* 21, no. 3 (2014), 606–35; Kathleen Coyne Doyle, "Thoreau's Flute, or, 'the Moose, the Pine Tree, & the Indian,' "*Concord Saunterer* 22 (2014), 99–111; Branka Arsić, *Bird Relics: Grief and Vitalism in Thoreau* (Cambridge, MA: Harvard University Press, 2016); and Melissa Sexton, "Sublime Matter: Materiality and Language in Thoreau's 'Ktaadn,'" in *Rediscovering the Maine Woods: Thoreau's Legacy in an Unsettled Land*, ed. John J. Kucich (Amherst: University of Massachusetts Press, 2019), 105–22.

9 Buell, *Environmental Imagination*, 116.

10 See John Aldrich Christie, *Thoreau as World Traveler* (New York: Columbia University Press and the American Geographical Society, 1965), 313–33.

11 See Robert Sattelmeyer, *Thoreau's Reading: A Study in Intellectual History with Bibliographic Catalog* (Princeton, NJ: Princeton University Press, 1988), 47–9.

12 Laura Dassow Walls, "Walking West, Gazing East: Planetarity on the Shores of Cape Cod," in *Thoreauvian Modernities: Transatlantic Conversations on an American Icon*, ed. François Specq, Laura Dassow Walls, and Michel Granger (Athens: University of Georgia Press, 2013), 21–42, 22.

13 Timothy Morton, *Ecology Without Nature: Rethinking Environmental Aesthetics* (Cambridge, MA: Harvard University Press, 2007), 30.

14 Buell, *The Environmental Imagination*, 84.

15 Ibid., 92.

16 Johnson's English dictionary cites uses of "excursion" by Pope, Locke, Arbuthnot, and Boyle, ranging from a deviation or ramble to an expedition, from "a progression beyond fixed limits" to a digression from some primary subject. See Samuel Johnson, *Johnson's English Dictionary* (London: Kimber & Sharpless, 1841), 357. Similarly, the *OED* charts the term's usages from the sixteenth century on, ranging from a deviation to a journey to a trip taken specifically for

pleasure. "excursion, n." *OED Online*, Oxford University Press (December 2020), www-oed-com.proxy.library.nd.edu/view/Entry/65949?rskey=vdJtB3&result=1.

17 On the title of this collection, see Joseph Moldenhauer's historical introduction in *Exc*, 342–4.

18 Wayne Franklin, *Discoverers, Explorers, Settlers: The Diligent Writers of Early America* (Chicago: University of Chicago Press, 1979), 16.

19 See Franklin, *Discoverers, Explorers, Settlers*, 59–68, 70–5.

20 On this change in travel writing and its significance for studies of Transcendentalism, and Thoreau in particular, see Lawrence Buell, *Literary Transcendentalism; Style and Vision in the American Renaissance* (Ithaca, NY: Cornell University Press, 1973), 188–207.

21 Laura Dassow Walls, *Henry David Thoreau: A Life* (Chicago: University of Chicago Press, 2017), 137.

22 Ibid., 158.

23 Linck C. Johnson, *Thoreau's Complex Weave: The Writing of A Week on the Concord and Merrimack Rivers* (Charlottesville: University Press of Virginia, 1986), xii–xiv.

24 Walls, *Henry David Thoreau*, 270.

25 Ibid., 268.

26 Johnson, *Thoreau's Complex Weave*, xvi.

27 See Tom Lynch, Cheryll Glotfelty, and KarlaArmbruster, eds., *The Bioregional Imagination: Literature, Ecology, and Place* (Athens: University of Georgia Press, 2012).

28 Thoreau would eventually survey the river as part of a major legal action initiated by Concord famers, whose hay meadows were eventually lost due to hydrological changes on the landscape. See Walls, *Henry David Thoreau*, 441–3. See also Robert Thorson, *The Boatman: Henry David Thoreau's River Years* (Cambridge, MA: Harvard University Press, 2019).

29 Buell, *Environmental Imagination*, 116–17.

30 In a letter to his cousin George Thatcher, Thoreau mentions a friend who "had neither woodcraft nor strength enough" for that ambitious trip. Thoreau, *Letters to a Spiritual Seeker*, 152.

31 Ibid., 158.

32 Ibid., 158–9.

33 Dieter Schulz, "Nature, Knowledge, and the Method of Thoreau's Excursions," in *Thoreauvian Modernities: Transatlantic Conversations on an American Icon*, ed. Francois Specq, Laura Dassow Walls, and Michel Granger (Athens: University of Georgia Press, 2013), 173.

34 Jeffrey Jerome Cohen and Lowell Duckert, "Welcome to the Whirled," in *Veer Ecology: A Companion for Environmental Thinking*, ed. Jeffrey Jerome Cohen and Lowell Duckert (Minneapolis: University of Minnesota Press, 2017), 2.
35 Nicholas Royle, *Veering: A Theory of Literature* (Edinburgh: Edinburgh University Press, 2011), 152, 153.
36 Cohen and Duckert, "Welcome to the Whirled," 3.
37 Survivance is, above all else, a tradition of "renunciations of dominance, tragedy, and victimry." Gerald Vizenor, *Manifest Manners: Narratives on Postindian Survivance* (Lincoln: University of Nebraska Press, 1999), vii.

1 "That West of Which I Speak": Imaginative Travel Writing

1 For centuries, Michilimackinac or Mackinac Island had been "the central point for all travel on the upper Great Lakes, and for a vast extent of wilderness and half-settled country beyond." Reuben Gold Thwaites, ed., *The Jesuit Relations and Allied Documents: Travels and Explorations of the Jesuit Missionaries in New France, 1610–1791; The Original French, Latin, and Italian Text, with English Translations and Notes*, vol. 55 (Cleveland, OH: Burrows Bros, 1896), 318n3. Fuller knew the island primarily by its Anglicized name, Mackinaw, though she writes that "its peculiar shape is supposed to have been the origin of the name, Michilimackinac which means the Great Turtle" (*Summer* 107). Today, it is usually spelled Mackinac, though the pronunciation is the same. In this chapter I use the longer name, Michilimackinac, to reflect the contested history of the region for Indigenous peoples as well as settlers, missionaries, and traders of French, British, and American origin. Despite Fuller's reflections on the etymology of the word, Michilimackinac is most likely the French transliteration of *Mishinimakinago*, the name "given to some strange Indians (according to the sayings of the Otchipwes [Ojibwe]), who are rowing through the woods, and who are sometimes heard shooting, but never seen" and "from this word, the name of the village of Mackinac, or Michillimackinac, is derived." Frederic Baraga, *A Dictionary of the Otchipwe Language*, vol. 2 (Montreal: Beauchemin & Valois, 1878), 248. See also Andrew Blackbird, *History of the Ottawa and Chippewa Indians of Michigan* (Ypsilanti, MI: Ypsilanti Auxiliary of the Woman's National Indian Association, 1887), 19–20.
2 For instance, Orestes Brownson complained in an 1844 review that *Summer on the Lakes* lacked "that tidiness which we look for in women," and Perry Miller, in a 1963 collection of Fuller's works, notes that the book "has been oddly slighted, even by devotees of 'the Margaret cult.'" Miller notes that the text is "a potpourri of a sort fashionable in the nineteenth century but tedious today." Orestes Brownson, Review

of *Summer on the Lakes*, in *Critical Essays on Margaret Fuller*, ed. Joel Myerson (Boston: G.K. Hall, 1980), 5–6; Perry Miller, *Margaret Fuller: American Romantic* (New York: Doubleday, 1963), 116. For overviews of this criticism, see Joel Myerson, *Margaret Fuller: A Descriptive Bibliography* (Pittsburgh, PA: University of Pittsburgh Press, 1978), xx–xxx; Charles Capper, *Margaret Fuller: An American Romantic Life*, vol. 2 (New York: Oxford University Press, 2007), 142–57.

3 James Freeman Clarke, Review of *Summer on the Lakes*, in *Critical Essays on Margaret Fuller*, ed. Joel Myerson (Boston: G.K. Hall, 1980), 3.

4 On Fuller's and Thoreau's influence on each other, see Marie Urbanski, "Henry David Thoreau and Margaret Fuller," *Thoreau Journal Quarterly* 8, no. 4, 1976, 24–30.

5 Irving went to Europe in 1815, and *The Sketch Book of Geoffrey Crayon, Gent.* was published serially in 1819 and 1820. Later, drawing from his experiences in Europe, he wrote *Bracebridge Hall, or The Humorists, A Medley* (1822), *Tales of a Traveller* (1824), and his Spanish books, *A History of the Life and Voyages of Christopher Columbus* (1828), *Chronicle of the Conquest of Granada* (1828), *Voyages and Discoveries of the Companions of Columbus* (1831), and *Tales of the Alhambra* (1832).

6 Stanley T. Williams and Barbara D. Simison, eds., Introduction to Henry Leavitt Ellsworth's *Washington Irving on the Prairie; or, A Narrative of the Tour of the Southwest in the Year 1832* (New York: American Book Company, 1937), xi.

7 On this clash of personal and professional interests, see Henry Leavitt Ellsworth, *Washington Irving on the Prairie; or, A Narrative of the Tour of the Southwest in the Year 1832*, ed. Williams and Simison (New York: American Book Company, 1937), 14.

8 Both Latrobe and Pourtalès would write their own accounts of the tour, and like Irving, both would attempt to capture their experiences through the rhetoric of a vanishing wilderness. See Charles Joseph Latrobe, *The Rambler in North America: 1832–33*, 2 vols. (London: R. B. Seeley and W. Burnside, 1835), esp. 1:6, 166–7, 203; and Albert-Alexandre Pourtalès, *On the Western Tour with Washington Irving: The Journal and Letters of Count de Pourtalès*, ed. George F. Spaulding (Norman: University of Oklahoma Press, 1968), esp. 19, 21–2. Ellsworth later offered his own account of the "*Irving* party" in his book-length letter to his wife, written several weeks after the trip. See Ellsworth, *Washington Irving on the Prairie*, esp. 14.

9 Lee Clark Mitchell, *Witnesses to a Vanishing America: the Nineteenth-Century Response* (Princeton, NJ: Princeton University Press, 1981), 26.

10 Ellsworth's journal, along with *Tour of the Prairies*, records multiple sexual encounters between Pourtalès and Native women. For example, see Ellsworth, *Washington Irving on the Prairie*, 14.
11 Mitchell, *Witnesses to a Vanishing America*, 26–7.
12 Qtd. in Washington Irving, *The Western Journals of Washington Irving*, ed. John Francis McDermott (Norman: University of Oklahoma Press, 1944), 10.
13 Gerald Vizenor, *Fugitive Poses: Native American Indian Scenes of Absence and Presence* (Lincoln: University of Nebraska Press, 1998), 108.
14 Ibid., 110.
15 Buell, *Literary Transcendentalism*, 198.
16 Peter Antelyes, *Tales of Adventurous Enterprise: Washington Irving and the Poetics of Western Expansion* (New York: Columbia University Press, 1990), 84.
17 Ibid., 96.
18 Wayne R. Kime, "The Author as Professional: Washington Irving's 'Rambling Anecdotes' of the West," in *Critical Essays on Washington Irving*, ed. Ralph M. Aderman (Boston: G.K. Hall, 1990), 241.
19 Mark K. Burns, "Ineffectual Chase: Indians, Prairies, Buffalo, and the Quest for the Authentic West in Washington Irving's A Tour on the Prairies," *Western American Literature* 42, no. 1 (Spring 2007), 58.
20 James D. Lilley, *Common Things: Romance and the Aesthetics of Belonging in Atlantic Modernity* (New York: Fordham University Press, 2014), 153.
21 Robert E. Abrams, *Landscape and Ideology in American Renaissance Literature: Topographies of Skepticism* (Cambridge: Cambridge University Press, 2004), 89–90.
22 For more on the Conversations and Fuller's influence on nineteenth-century oral culture, see Judith Matteson Bean, "'A *Presence* Among Us': Fuller's Place in Nineteenth-Century Oral Culture," *ESQ: A Journal of the American Renaissance* 44, nos. 1–2 (1998); and Charles Capper, "Margaret Fuller as Cultural Reformer: The Conversations in Boston," *American Quarterly* 39, no. 4 (1987).
23 Caroline W. Healey Dall, *Margaret and Her Friends* (Boston: Roberts Brothers, 1895), 13. See also Megan Marshall, *Margaret Fuller: A New American Life* (Boston: Houghton Mifflin Harcourt, 2013), 164–8. For the publication history of Healey's account of the series, see Joel Myerson, "Mrs. Dall Edits Miss Fuller: The Story of Margaret and Her Friends," *Papers of the Bibliographic Society of America* 72, no. 2 (1978), 187–200.
24 "The Great Lawsuit" was published by Emerson in the July 1843 issue of *The Dial*, while Fuller was traveling in Illinois. In 1845 she published a revised and expanded version as a book, *Woman in the Nineteenth Century*. For a reading of the essay and the ways that it moderates Transcendentalist aesthetics and feminism,

see Sandra Gustafson, "Choosing a Medium: Margaret Fuller and the Forms of Sentiment," *American Quarterly* 47, no. 1 (March 1995). Also see Phyllis Cole, "The Nineteenth-Century Women's Rights Movement and the Canonization of Margaret Fuller," *ESQ: A Journal of the American Renaissance* 44, nos. 1–2 (1998).

25 Fuller had also translated Eckermann's *Conversations with Goethe*, which was published in the United States in 1839. See Karen English, "Genuine Transcripts of Private Experience," *American Transcendental Quarterly* 15, no. 2 (2001); and Elisabeth Lenckos, " 'Stimulus and Cheer': Margaret Fuller's 'Translations,' from Eckermann's Conversations with Goethe to Bettina von Arnim's Guenderode," in *Translators, Interpreters, Mediators: Women Writers 1700–1900*, ed. Gillian E. Dow (Bern: Peter Lang, 2007).

26 At Emerson's suggestion, Carlyle published his books simultaneously in England and the United States to avoid the inevitable pirating and illegal sale of his work on American markets. See Chris R. Vanden Bossche, "Notes on the Text," in *Past and Present*, ed. Thomas Carlyle (Berkeley: University of California Press, 2005).

27 Margaret Fuller, *The Letters of Margaret Fuller*, ed. Robert N. Hudspeth, vol. 3 (Ithaca, NY: Cornell University Press, 1984), 128.

28 Lance Newman, *Our Common Dwelling: Henry Thoreau, Transcendentalism, and the Class Politics of Nature* (New York: Palgrave Macmillan, 2005), 61.

29 Annette Kolodny, *The Land Before Her: Fantasy and Experience of the American Frontiers, 1630–1860* (Charlottesville: University of North Carolina Press, 1984), 129.

30 Newman, *Our Common Dwelling*, 67.

31 In an 1840 letter to William H. Channing, Fuller writes that "Utopia is impossible to build up. At least, my hopes for our race on this one planet are more limited than those of most of my friends. I accept the limitations of human nature, and believe a wise acknowledgement of them one of the best conditions of progress." Margaret Fuller, *The Letters of Margaret Fuller*, vol. 2, ed. Robert N. Hudspeth (Ithaca, NY: Cornell University Press, 1983), 109. See also Marshall, *Margaret Fuller*, 186–7, 331.

32 See Lauren Berlant, *Cruel Optimism* (Durham, NC: Duke University Press, 2011), 4ff.

33 Newman, *Our Common Dwelling*, 65.

34 Kolodny, *The Land Before Her*, 128–9.

35 William W. Stowe, "Conventions and Voices in Margaret Fuller's Travel Writing," *American Literature* 63, no. 2 (June 1991), 256.

36 Margaret V. Allen, *The Achievement of Margaret Fuller* (University Park: Pennsylvania State University Press, 1979), 561.

37 Newman, *Our Common Dwelling*, 65.

38 The distinction in this sentence between Gothic and Roman modes of settlement establishes a crucial contrast between what Fuller sees as the unmediated violence of an invading horde and, by contrast, the orderly settlement of republican citizens. See Newman, *Our Common Dwelling*, 65.

39 Charles Capper offers a helpful analysis of this ideology, noting that it would actually "be more accurate to call it three savagisms, each with its objectified Other: the popular racist's bloodthirsty animal bent on killing whites; the Enlightenment's anachronism outside 'civilization' based on private property, sociability, and Christian virtue; and the Romantic 'Noble Savage,' independent, eloquent, and stoic." Charles Capper, *Margaret Fuller: An American Romantic Life*, vol. 2 (Oxford: Oxford University Press, 2007), 148. See also Roy Harvey Pearce, *Savages and Civilization: A Study of the History of the Indian and the American Mind* (Berkeley: University of California Press, 1988; originally published 1953); Robert F. Berkhofer, Jr., *The White Man's Indian: Images of the American Indian from Columbus to the Present* (New York: Knopf, 1978); Brian W. Dippie, *The Vanishing American: White Attitudes and the U.S. Indian Policy* (Berkeley: University of California Press, 1982); Philip Deloria, *Playing Indian* (New Haven, CT: Yale University Press, 1998); and Steve Conn, *History's Shadow: Native Americans and Historical Consciousness in the Nineteenth Century* (Chicago: University of Chicago Press, 2004).

40 Despite these explicit denunciations of savagism, some scholars have argued that Fuller's rhetoric evinces a symbolic erasure of Native peoples. See Lucy Maddox, *Removals: Nineteenth-Century American Literature and the Politics of Indian Affairs* (Oxford: Oxford University Press, 1991), 139–48; and Nicole Tonkovich, "Traveling in the West and Writing in the Library: Margaret Fuller's *Summer on the Lakes*," *Legacy* 10 (1993), 79–102. Contrasting interpretations include Christina Zwarg, *Feminist Conversations: Fuller, Emerson, and the Play of Reading*, (Ithaca, NY: Cornell University Press, 1995), ch. 3; and Charlene Avallone, "The Red Roots of White Feminism in Margaret Fuller's Writings," in *Doing Feminism: Teaching and Research in the Academy*, ed. Mary Anderson, Lisa Fine, Kathleen Geissler, and Joyce R. Landenson (East Lansing: Michigan University Press, 1997), 135–65.

41 Susan Belasco Smith, "Introduction," in *Summer on the Lakes*, ed. Margaret Fuller (Champaign: University of Illinois Press, 1991), xv.

42 Cheryl J. Fish, *Black and White Women's Travel Narratives* (Gainesville: University of Florida Press, 2004), 100.

43 Charles Capper, *Margaret Fuller: An American Romantic Life*, vol. 2 (Oxford: Oxford University Press, 2007), 142.

44 Fish, *Black and White Women's Travel Narratives*, 96.

45 Tonkovich, "Traveling in the West, Writing in the Library," 80.

46 See Philip F. Gura, *American Transcendentalism: A History* (New York: Hill and Wang, 2007), 226. See also Catherine C. Mitchell, ed., *Margaret Fuller's New York Journalism: A Biographical Essay and Key Writings* (Knoxville: University of Kentucky Press, 1995), 3–50.
47 Gura, *American Transcendentalism*, 227.
48 See Laura Dassow Walls, *Henry David Thoreau: A Life* (Chicago: University of Chicago Press, 2017), 255.

2 "Some Foreign Vortex": Disturbance and the Limits of Genre

1 While preparing for these lectures, Thoreau composed a draft "Course of Lectures on Cape Cod," which he later revised and interlined with new details from the June 1850 trip (Manuscript at the Huntington Library, San Marino, California, HM 13206). See Moldenhauer, "Historical Introduction," in *CC* 253–4.
2 "Yankee in Canada" first appeared in *Putnam's* under the title "Excursion to Canada." The piece was then collected in *Yankee in Canada, with Anti-Slavery and Reform Papers* (1866). See Moldenhauer in *Exc* 497n79.1.
3 Thoreau continued working on *Cape Cod* until at least 1860, and he made limited changes to "Yankee in Canada," as well. See Robert Sattelmeyer, *Thoreau's Reading: A Study in Intellectual History with Bibliographical Catalogue* (Princeton, NJ: Princeton University Press, 1988), 92–9; Moldenhauer, "Historical Introduction," in *CC* 267–85; and Moldenhauer, "Headnote," in *Exc* 478–82.
4 In an unpublished notebook, the two projects commingle and mirror each other, the notes for "Yankee in Canada" occupying the first pages and those for *Cape Cod* written upside down at the back of the same volume. This notebook also marks the beginning of Thoreau's unpublished Indian Books, which record a decade of research at Harvard Library and beyond. See Linck Johnson, "Into History: Thoreau's Earliest 'Indian Book' and the His First Trip to Cape Cod," *Emerson Society Quarterly* 107 (1982). For a reproduction of this notebook in Thoreau's hand, without transcription or editorial commentary, see Kenneth Walter Cameron, ed., *Transcendental Climate; New Resources for the Study of Emerson, Thoreau and Their Contemporaries* (Hartford, CT: Transcendental Books, 1963), 310–411. The notebook was also reproduced as a standalone volume. See Cameron, *Thoreau's Canadian Notebook, and Record of Surveys: Selected Chapters from Transcendental Climate* (Hartford, CT: Transcendental Books, 1967).
5 Greeley's sense of the "good time" refers to the "June Nos. of the Monthlies," at the beginning of the summer tourist season, when Americans would be turning

their attention toward vacation destinations in earnest. The phrasing also suggests that the setting would need to be fresh. The summer before, Charles A. Dana had traveled from New York to Montreal, Quebec, and the Saguenay, some 100 miles further down the St. Lawrence from St. Anne. His series of five letters recounting the trip appeared in Greeley's own *New York Tribune* from August 29 to September 26, 1851.

6 Thoreau sent the *Cape Cod* manuscript to Curtis on November 16, 1852. A week later, on November 23, Greeley wrote defensively: "I have made no bargain—none whatever—with Putnam, concerning your MS. I have indicated no price to him. I handed over the MS. because I wish it published, and presumed that was in accordance both with your interest and your wishes." He goes on: "Your 'Canada' is not so fresh and acceptable as if it had just been written on the strength of a last summer's trip, and I hope you will have it printed in Putnam's Monthly. But I have said nothing to his folks, as to price, and will not till I hear from you again" (*Corr* 2:133).

7 Moldenhauer, "Headnote," in *Exc* 479.

8 Robert D. Richardson, Jr., *Henry Thoreau: A Life of the Mind* (Berkeley: University of California Press, 1986), 216.

9 John Aldrich Christie, *Thoreau as World Traveler* (New York: Columbia University Press, 1965), 100.

10 Walter Harding, *A Thoreau Handbook* (New York: New York University Press, 1959), 47. Harding echoes this line in his subsequent biography, calling "Yankee in Canada" "the least successful" of the excursions. See Harding, *The Days of Henry Thoreau: A Biography* (Princeton, NJ: Princeton University Press, 1992), 283.

11 See Joseph Lawrence Basile, "Thoreau's Uncommon Cold: 'A Yankee in Canada' Revisited," *Concord Saunterer* 13, no. 1 (1978), 7–11.

12 Stephen Adams, "Thoreau Catching Cold: 'A Yankee in Canada,'" *Emerson Society Quarterly* 25, no. 97 (1979), 225.

13 Ibid., 224, 226.

14 Patrick Lacroix, "Finding Thoreau in French Canada: The Ideological Legacy of the American Revolution," *American Review of Canadian Studies* 47, no. 3 (2017), 267.

15 Walls, *Henry David Thoreau*, 300.

16 In invoking *an* end of the world, rather than *the* end, I am following Kate Rigby's suggestion that an apocalypse might productively be read as "the end of a particular practice, or set of practices of world-making," and even "a trigger for transformation" rather than a simple endpoint or an irrevocable catastrophe. Apocalypse can even be a kind of *revelation.* Kate Rigby, *Dancing with Disaster: Environmental Histories, Narratives, and Ethics for Perilous Times* (Charlottesville: University of Virginia Press, 2015), 38.

17 For more on the environmental history of Concord prior to the rise of the railroad and the development of national and global markets, see Brian Donahue, *The Great Meadow: Farmers and the Land in Colonial Concord* (New Haven, CT: Yale University Press, 2004). For broader accounts of the environmental history of New England, see Richard Judd, *Second Nature: An Environmental History of New England* (Amherst: University of Massachusetts Press, 2014); and William Cronon, *Changes in the Land: Indians, Colonists, and the Ecology of New England* (New York: Hill and Wang, 1983).

18 The St. Lawrence Valley contains a number of trees, shrubs, and plants that Thoreau would have recognized as native to Maine and even to Concord—in this chapter alone he catalogs wild apples, spruce, firs, birches, aspens, arbor vitae, beeches, maples, basswood, and white pines, among other trees, noting the absence only of New England's ubiquitous pitch pines. Furthermore, peas and beans, he notes, are important crops in the region, and are currently being harvested, along with some less familiar crops, *snells* and "pinbéna" (*Exc* 117–18, 122).

19 Sara Ahmed, "Happy Objects," in *Affect Theory Reader*, ed. Gregory Seigworth and Melissa Gregg (Durham, NC: Duke University Press, 2010), 29.

20 My thinking on local places and their connections to distant geographies around the planet is influenced by Tom Lynch, Susan Naramore Maher, Drucilla Wall, and O. Alan Weltzien, "An Alignment of Stones," in *The Bioregional Imagination: Literature, Ecology, and Place*, ed. Tom Lynch, Susan Naramore Maher, Drucilla Wall, and O. Alan Weltzien (Lincoln: University of Nebraska Press, 2017), xi–xxvi.

21 Timothy Clark, *The Cambridge Introduction to Literature and Environment* (Cambridge: Cambridge University Press, 2011), 197.

22 Ibid.

23 John Hessler, *Henry David Thoreau on Cape Cod* (Delray Beach, FL: Levenger Press, 2010), 222.

24 Kensett, a second-generation member of the Hudson River School, painted in both the "Grand Style" evocative of the sublime, often associated with work by Thomas Cole, and a "transcendental luminism," characterized by understatement, stillness, and a quiet, contemplative atmosphere. See Barbara Novack, *Nature and Culture: American Landscape and Painting, 1825–1875* (Oxford: Oxford University Press, 1980), 28–9. For a brief interpretation of *Lotus-Eating* focusing on Kensett's illustrations, see Melissa Trafton, "'It is a Joint Venture': John Frederick Kensett's Images for *Lotus-Eating*," *American Art* 25, no. 2 (2011), 105–8.

25 On this gradual change in American travel writing and the means by which Transcendentalists contributed to and challenged the genre, see Buell, *Literary Transcendentalism*, 188–207.

26 Qtd. in Gordon Milne, *George William Curtis & the Genteel Tradition* (Bloomington: Indiana University Press, 1956), 17.

27 See Milne, *George William Curtis & the Genteel Tradition*, 30n11. See also Walls, *Henry David Thoreau*, 188.

28 Buell, *Literary Transcendentalism*, 191.

29 Bayard Taylor's poetry, too, was influenced by travel writing. His first book, which included "Ximena, or the Battle of the Sierra Morena" and fourteen other poems, was published in 1844 with an epigraph by the English poet Henry Kirk White: "I am a youthful traveler in the way."

30 James Weaver, "The Continent Behind: Alienation and the American Scene in George William Curtis's *Lotus-Eating: A Summer Book*," *Journeys* 17, no. 1 (2016), 30.

31 Ibid. For brief accounts of Curtis's contribution to American tourism and travel writing, see John F. Sears, *Sacred Places: American Tourist Attractions in the Nineteenth Century* (New York: Oxford University Press, 1989), 60–1, 63; Hans Huth, *Nature and the American: Three Centuries of Changing Attitudes* (Lincoln: University of Nebraska Press, 1957), 83–4; and Edward Halsey Foster, *The Civilized Wilderness: Backgrounds to American Romantic Literature, 1817–1860* (New York: Free Press, 1975), 15–16.

32 See James Schlett, *A Not Too Greatly Changed Eden: The Story of the Philosophers' Camp in the Adirondacks* (Ithaca, NY: Cornell University Press, 2015).

33 Weaver, "The Continent Behind," 30.

34 Joseph Rodman Drake's *The Culprit Fay and Other Poems* was published posthumously in 1835. Edgar Allen Poe, reviewing the book in 1836 in the *Southern Literary Messenger*, criticized it for requiring that readers "imagine a race of Fairies in the vicinity of West Point." See Poe, *Essays and Reviews* (New York: Library of America), 521.

35 Thoreau, *Letters to a Spiritual Seeker*, 158.

3 Affect and Ecological Form in *Walden* and *Cape Cod*

1 See John Aldrich Christie, *Thoreau as World Traveler* (New York: Columbia University Press and the American Geographical Society, 1965), 3–41.

2 Laura Dassow Walls, *Henry David Thoreau: A Life* (Chicago: University of Chicago Press, 2017), 273.

3 Linck C. Johnson, *Thoreau's Complex Weave: The Writing of A Week on the Concord and Merrimack Rivers* (Charlottesville: University Press of Virginia, 1986), 270. In

May 1848, Thoreau wrote to Greeley, "My book is swelling again under my hands" (*Corr* 1:376). See also Walls, *Henry David Thoreau*, 254–64.

4 See J. Lyndon Shanley, *The Making of Walden, with the Text of the First Edition* (Chicago: University of Chicago Press, 1957), 30–1. After Shanley published the text of Thoreau's first draft, Ronald Earl Clapper compiled a complete edition of the manuscript and its various revisions across these drafts, the corrected version of which forms the basis of the Digital Thoreau "fluid text" edition of *Walden*. Ronald Earl Clapper, "The Development of Walden: A Genetic Text" (PhD dissertation, University of California, 1967), and Henry David Thoreau, *Walden: A Fluid-Text Edition* (Digital Thoreau, https://digitalthoreau.org/fluid-text-toc).

5 William Rossi, "Making *Walden* and Its Sandbank," *Concord Saunterer* 30 (2022), 10–58, 20.

6 Laura Dassow Walls argues that Thoreau's "turn" was influenced by his reading of Alexander von Humboldt, whose "empirical holism" offered a means of seeing new worlds, while Robert Thorson marks Thoreau's reading of Darwin's *Journal of Researches* (1839) as a pivotal moment. Laura Dassow Walls, *Seeing New Worlds: Henry David Thoreau and Nineteenth-Century Natural Science* (Madison: University of Wisconsin Press, 1995), 94–130; and Robert Thorson, *Walden's Shore: Henry David Thoreau and Nineteenth-Century Science* (Cambridge, MA: Harvard University Press, 2014), 121–3. See also Walls, *Henry David Thoreau*, 273–312.

7 James Perrin Warren, *Thoreau's Botany: Thinking and Writing with Plants* (Charlottesville: University of Virginia Press, 2023), 16.

8 Abram coined the term "more-than-human world" to encompass "the larger community," which includes, "along with the humans, the multiple nonhuman entities that constitute the local landscape, from the diverse plants and the myriad animals—birds, mammals, fish, reptiles, insects—that inhabit or migrate through the region, to the particular winds and weather patterns that inform the local geography, as well as the various landforms—forests, rivers, caves, mountains—that lend their specific character to the surrounding earth." David Abram, *The Spell of the Sensuous: Perception and Language in a More-Than-Human World* (New York: Vintage Books, 1996), 6–7.

9 Lawrence Buell, *The Environmental Imagination: Thoreau, Nature Writing, and the Formation of American Culture* (Cambridge, MA: Harvard University Press, 1995), 138. See also 116–26.

10 Robert Sattelmeyer, *Thoreau's Reading: A Study in Intellectual History with Bibliographical Catalog* (Princeton, NJ: Princeton University Press, 1988), 48.

11 Christie, *Thoreau as World Traveler*, 267.

12 Buell, *Environmental Imagination*, 116.

13 Laura Dassow Walls, "Walking West, Gazing East: Planetarity on the Shores of Cape Cod," in *Thoreauvian Modernities: Transatlantic Conversations on an American Icon*, ed. François Specq, Laura Dassow Walls, and Michel Granger (Athens: University of Georgia Press, 2013), 21–42, 22.

14 On Freeman and others like him who appear in *Walden*, see Elise Lemire, *Black Walden: Slavery and Its Aftermath in Concord, Massachusetts* (Philadelphia: University of Pennsylvania Press, 2009).

15 Robert D. Richardson, *Henry Thoreau: A Life of the Mind* (Oakland: University of California Press, 1986), 256. For a summary of earlier interpretations of this passage, see Gordon V. Boudreau, *The Roots of Walden and the Tree of Life* (Nashville, TN: Vanderbilt University Press, 1990), 117–34. More recent research focusing on this passage includes Thomas Peyser, "*Walden's* Sandbank and the *Vestiges* Controversy," *American Notes and Queries* 30, no. 1 (2017), 126–36 and Rossi, "Making *Walden* and Its Sandbank."

16 For a technical scientific summary of this phenomena and of Thoreau's description, see Thorson, *Walden's Shore*, 283.

17 See Warren, *Thoreau's Botany*, 81–91.

18 Walls, *Henry David Thoreau*, 355.

19 Richardson, *Henry Thoreau*, 256, 312.

20 Buell, *Environmental Imagination*, 125. Buell notes that the late natural history essay "Huckleberries" is perhaps a better example of Thoreau's most mature environmental thinking.

21 Rossi, "Making *Walden* and Its Sandbank," 24. For the original text and revisions of the "core," see Rossi, "Making *Walden* and Its Sandbank," 30–43. See also *PJ* 2:382–4 and 2:587–9.

22 Rossi, "Making *Walden* and Its Sandbank," 25.

23 Walls, *Henry David Thoreau*, 354.

24 On *Cape Cod* as a "sunny" essay celebrating New England culture, see Walter Roy Harding, *The Days of Henry Thoreau* (New York: Knopf, 1967), 361. For early scholarship on the darker characteristics of *Cape Cod*, see, for example, Richard Bridgman, *Dark Thoreau* (Lincoln: University of Nebraska Press, 1982), 158–87; Robert Pinsky, "Comedy, Cruelty, and Tourism: Thoreau's *Cape Cod*," *American Scholar* 79, no. 3 (2004), 79–88; Mitchell Breitwieser, *National Melancholy: Mourning and Opportunity in Classic American Literature* (Stanford, CA: Stanford University Press, 2007), 144–59; and Ryan Schneider, "Drowning the Irish: Natural Borders and Class Boundaries in Henry David Thoreau's *Cape Cod*," *American Transcendental Quarterly* 22, no. 3 (2008), 463–76.

25 Katie Simon, "Affect and Cruelty in the Atlantic System: The Hauntological Argument of Henry David Thoreau's *Cape Cod*," *ESQ: A Journal of*

Nineteenth-Century Literature and Culture 62, no. 2 (2016), 245–82, 245. See also Lawrence Buell, "Henry Thoreau Enters the American Canon," in *New Essays on Walden*, ed. Robert F. Sayre (Cambridge: Cambridge University Press, 1992), 23–55.

26 Simon, "Affect and Cruelty in the Atlantic System," 246.

27 Pinsky, "Comedy, Cruelty, and Tourism," 281–2.

28 Alex Moskowitz, "Economic Imperception; or, Reading Capital on the Beach with Thoreau," *American Literary History* 32, no. 2 (2020), 221–42, 221. See also Benjamin Pickford, "*Cape Cod*, Literature, and the Illocality of Thinking About Capital" in *Thoreau Beyond Borders*, ed. François Specq, Laura Dassow Walls, and Julien Nègre (Amherst: University of Massachusetts Press, 2020), 179–94. On the wreck of the *St. John* and social class, see Schneider, "Drowning the Irish," 463–76.

29 Jamie L. Jones, "Long Looking, Without Seeing: Thoreau's *Cape Cod* and Global Capitalism at Sea," *The Dial: A Journal of the Ralph Waldo Emerson Society* 1, no. 1 (2024), 42–65, 55.

30 Paul Giles, "Transnational Thoreau: Time, Space, and Relativity," in *Thoreau at 200: Essays and Reassessments*, ed. Kristen Case and K. P. Van Anglen (Cambridge: Cambridge University Press, 2016), 138–53, 150.

31 Christa Holm Vogelius, "*Cape Cod*'s Transnational Bodies," in *Thoreau Beyond Borders*, ed. François Specq, Laura Dassow Walls, and Julien Nègre (Amherst: University of Massachusetts Press, 2020), 25–39, 27. Laura Dassow Walls also compares *Cape Cod* to "Walking." See Laura Dassow Walls, "Walking West, Gazing East: Planetarity on the Shores of Cape Cod," in *Thoreauvian Modernities: Transatlantic Conversations on an American Icon*, ed. François Specq, Laura Dassow Walls, and Michel Granger (Athens: University of Georgia Press, 2013), 21–42.

32 For example, see John Hessler, "From Ortelius to Champlain: The Lost Maps of Henry David Thoreau," *Concord Saunterer* 18 (2010), 1–26; Michael C. Weisenberg, "Beyond the Borders of Time: Thoreau and the 'Ante-Pilgrim History' of the New World," in *Thoreau Beyond Borders*, ed. François Specq, Laura Dassow Walls, and Julien Nègre (Amherst: University of Massachusetts Press, 2020), 40–54; and Benjamin Vilhauer, "The Theme of Time in Thoreau's *Cape Cod*: The Crisis of the Present Shore," *Concord Saunterer* 16 (2008), 33–44.

33 On Thoreau's reading during this period, see Moldenhauer, "Historical Introduction," in *CC* 251–3. See also Sattelmeyer, *Thoreau's Reading*, 54–77.

34 Joseph Moldenhauer, "Historical Introduction," in *CC* 256. Thoreau draws from his experiences on Fire Island in "The Beach Again," *CC* 84–5. On the effects of this trip on the narrative and on Thoreau's writing and thinking, see Vogelius, "*Cape Cod*'s Transnational Bodies."

35 This anecdote was added later, after 1857. See Moldenhauer, "Historical Introduction" in *CC* 282–3.
36 Moldenhauer, "Historical Introduction," in *CC* 266.
37 Warren notes that Thoreau's descriptions of botanical observations, mostly from this third trip to the Cape, are an essential difference between the first and second halves of the book. Warren, *Thoreau's Botany*, 58.
38 See *Corr* 2:339–40.
39 Moldenhauer, "Historical Introduction," in *CC* 276.
40 According to Moldenhauer, the 1856 visit offered an introduction to menhaden and an anecdote about quahogs, which appears in "The Wellfleet Oysterman." The 1857 visit, meanwhile, records an experiment making tallow from bayberries, which Thoreau recorded in his journal on April 7 and 8 and which was ultimately included in some detail in "The Beach Again." See Moldenhauer, "Historical Introduction," in *CC* 280–1.
41 This omitted material is nonetheless, per Warren, a key example of Thoreau's botanizing as integral to his mature excursion form. See Warren, *Thoreau's Botany*, 63–8.
42 On these later additions and revisions, see Moldenhauer, "Historical Introduction," 278–85.
43 Walls, "Walking West, Gazing East," 34.
44 "Highland Light," Cape Cod National Seashore, National Park Service, 2023, www.nps.gov/places/highland-light.htm.
45 Weisenberg, "Beyond the Borders of Time," 41.

4 George Copway's Figures of Homesickness

1 "Ojibwe" is a broad term for several loosely affiliated groups of Native peoples living in the Great Lakes, on the Canadian prairies, and in the upper reaches of the Mississippi River basin, all of whom recognized distinct cultural identities. These groups have been known by various names throughout their territory, including the Mississaugas or Mississauga Ojibwe in southern Ontario, the Saulteaux in western Canada, and the Chippewas in the United States. Throughout their range, they generally refer to themselves as Anishinaabe (variously spelled), which loosely translates to "first people." In this chapter I use the most common English spelling "Ojibwe," preserving Copway's spellings, usually "Ojebwa," in quotations from his writing. For histories of the Ojibwe peoples, see Thomas D. Peacock, Marlene Wisuri, and Winona LaDuke, *Ojibwe Waasa Inaabidaa: We Look in All Directions* (Afton, MN: Afton Historical Society Press, 2001); and Peter S. Schmalz,

The Ojibwa of Southern Ontario (Toronto: University of Toronto Press, 1991). Nineteenth-century histories of the Ojibwe written by Native authors include Peter Jones, *History of the Ojebway Indians: With Especial Reference to Their Conversion to Christianity* (London: A. W Bennett, 1861); and Copway's own *Traditional History and Characteristic Sketches of the Ojibway Nation* (London: C. Gilpin, 1850). William W. Warren, of mixed Ojibwe-Metis and white descent, also wrote a nineteenth-century account of the Ojibwe: *History of the Ojibway People* (St. Paul: Minnesota Historical Society Press, 1984, originally published 1885).

2 Copway's autobiography was first published as *Life, History, and Travels of Kah-ge-ga-gah-bowh (George Copway)* (1847). A slightly revised edition, which appended speeches and published letters, was published in 1850 under two titles: in New York as *The Life, Letters and Speeches of Kah-ge-ga-gah-bowh or, G. Copway, Chief Ojibway Nation*, and in London as *Recollections of a Forest Life: or, the Life and Travels of Kah-ge-ga-gah-bowh, or George Copway, Chief of the Ojibwey Nation.* This chapter refers to the text by its earliest title, *Life, History, and Travels.*

3 Cathy Rex, "Survivance and Fluidity: George Copway's *The Life, History, and Travels of Kah-ge-ga-gah-bowh*," *Studies in American Indian Literatures* 18, no. 2 (2006), 5.

4 William Whipple Warren, born in La Pointe in 1825, perhaps reflecting on people like Copway, wrote, "There are several villages of Indians in Upper Canada, who are sometimes denominated as Ojibways … . The connection existing between these and the Lake Superior and Mississippi Ojibways, is not very close, though they speak the same language, and call each other relatives." Warren, *History of the Ojibwe People*, 368.

5 Sense of place refers to individual and cultural relationships between people and geographies. See Buell, *The Future of Environmental Criticism: Environmental Crisis and Literary Imagination* (Malden, MA: Blackwell, 2005), 62–96. See also Yi Fu Tuan, *Space and Place: The Perspective of Experience* (Minneapolis: University of Minnesota Press, 1977); and Erica Carter, James Donald, and Judith Squires, eds., *Space and Place: Theories of Identity and Location* (London: Lawrence and Wishart, 1993).

6 Glenn Albrecht, Gina-Maree Sartore, Linda Connor, Nick Higginbotham, Sonia Freeman, Brian Kelly, Helen Stain, Anne Tonna, and Georgia Pollard, "Solastalgia: The Distress Caused by Environmental Change," *Australian and New Zealand Journal of Psychiatry* 15, no. S1 (2007), S96.

7 Like many authors of the nineteenth century, Copway regularly borrowed material from other published writing, particularly in *Running Sketches* and *Organization of a New Indian Territory.* His biography was likely edited by his wife Elizabeth, a white woman, but scholars agree that *Life, History, and Travels* was his own work.

See Donald B. Smith, "Kahgegagahbowh: Canada's First Literary Celebrity in the United States," in *Life*, 34–5.

8 John J. Kucich and Caitlin Cribben argue that these annotations reflect a "missing Indian notebook," reading Thoreau's marginalia as part of his extensive research on Native peoples. John J. Kucich and Caitlin Cribben, "Thoreau's Reading of George Copway: An Additional Indian Notebook," *Thoreau Society Bulletin*, no. 302 (Summer 2018), 1.

9 Rex, "Survivance and Fluidity," 2.

10 Bernd C. Peyer, *The Tutor'd Mind: Indian Missionary-Writers in Antebellum America* (Amherst: University of Massachusetts Press, 1997), 236.

11 Peyer, *The Tutor'd Mind*, 224.

12 Donald B. Smith, "The Life of George Copway," *Journal of Canadian Studies* 23, no. 3 (Fall 1988), 25, 29.

13 Smith, "Kahgegagahbowh," 32–3.

14 Maureen Konkle, *Writing Indian Nations: Native Intellectuals and the Politics of Historiography, 1827–1863* (Chapel Hill: University of North Carolina Press, 2004), 33.

15 Robert Allen Warrior, *Tribal Secrets: Recovering American Indian Intellectual Traditions* (Minneapolis: University of Minnesota Press, 1995), xix.

16 Scott Richard Lyons, *X-Marks: Native Signatures of Assent* (Minneapolis: University of Minnesota Press, 2010), 3.

17 Copway references this migration within *Life, History, and Travels*, where he proposes traveling in the western portions of the Ojibwe territory specifically to record the story in greater detail (89).

18 Lyons, *X-Marks*, 2.

19 The most thorough history of this region's complex history to focus on national allegiances and Indigenous groups is Michael Witgen's *An Infinity of Nations: How the Native New World Shaped Early North America* (Philadelphia: University of Pennsylvania Press, 2012).

20 Lyons, *X-Marks*, 2.

21 Konkle, *Writing Indian Nations*, 6.

22 Lyons, *X-Marks*, 12. Lyons is not alone in calling for a certain Native nationalism based on specific tribal traditions and cultures. In particular, see Craig S. Womack, *Red on Red: Native American Literary Separatism* (Minneapolis: University of Minnesota Press, 1999).

23 Charles J. Kappler, ed., *Indian Affairs: Laws and Treaties*, vol. II (Washington, DC: Government Printing Office, 1904), 542–5.

24 Article III of the Treaty of 1842 mentions removal directly: "It is agreed by the parties of this treaty, that whenever the Indians shall be required to remove from

the ceded district, all the unceded lands belonging to the Indians of Fond du Lac, Sandy Lake, and Mississippi bands, shall be the common property and home of all the Indians, property to this treaty." Kappler, *Indian Affairs*, 543. On Copway's awareness of removal, see Donald B. Smith, *Mississauga Portraits: Ojibwe Voices from Nineteenth-Century Canada* (Toronto: University of Toronto Press, 2013), esp. 178–9.

25 Survivance entails the various ways in which Indigenous peoples resist "dominance, tragedy, and victimry." See Gerald Vizenor, *Manifest Manners: Narratives on Postindian Survivance* (Lincoln: University of Nebraska Press, 1999).

26 On this point, see Konkle, *Writing Indian Nations*, 192.

27 For a history of this criticism, see A. LaVonne Brown Ruoff, "The Literary and Methodist Contexts of George Copway's *Life, Letters, and Speeches*," in *Life*, 1–21.

28 Barry O'Connell, ed., *On Our Own Ground: The Complete Writings of William Apess, A Pequot* (Amherst: University of Massachusetts Press, 1992), 1–2. See also Ruoff, "Literary and Methodist Contexts," 2–3.

29 Copway invokes his father in a fit of homesickness just outside of La Pointe, Wisconsin, early in his career as a missionary (*Life* 112). He is mentioned again only at the end of the book, in "A Geographical Sketch of the Ojebwa, or Chippeway Nation," where Copway merely lists the surnames of three chiefs of the "Chippewas at Rice Lake" (*Life* 157).

30 Hertha D. Sweet Wong, "Native American Life Writing," in *Cambridge Companion to Native American Literature*, ed. Joy Porter and Kenneth M. Roemer (Cambridge: Cambridge University Press, 2005), 133.

31 William Bevis, "Native American Novels: Homing In," in *Recovering the Word: Essays on Native American Literature* (Oakland: University of California Press, 1987), 582.

32 My thinking here has been influenced by Richard Scott Lyons, who argues that Copway's *Running Sketches of Men and Places, in England, France, Germany, Belgium, and Scotland* similarly reflects the conventions of nineteenth-century transatlantic travel writing. Lyons, "Migrations to Modernity: The Many Voices of George Copway's *Running Sketches of Men and Places, in England, France, Germany, Belgium, and Scotland*," in *The World, the Text, and the Indian: Global Dimensions of Native American Literature*, ed. Scott Richard Lyons (New York: SUNY Press, 2017).

33 See "Ceded Territory Boundary" (Treaties of 1836, 1837, 1842 and 1854) [Map] (Odanah, WI: Great Lakes Indian Fish and Wildlife Commission, January 2015), https://data.glifwc.org/ceded/.

34 Charles J. Kappler, ed., *Indian Affairs: Laws and Treaties*, vol. II (Washington, DC: Government Printing Office, 1904), 542–5.

35 Vizenor, *Manifest Manners*, 12, 14.

36 In claiming that Copway finds himself in the Anthropocene, I do not mean to date the onset of the present ecological crisis or—worse—to implicate Copway. Similarly, I do not mean to suggest that he was aware of the Anthropocene as a looming concept. Rather, I mean to draw attention to the ways that Copway responded to the rise of modernity and new forms of travel during the nineteenth century, and the ways that the very form of his writing reflects that awareness.

37 Jonathan Lear, *Radical Hope: Ethics in the Face of Cultural Devastation* (Cambridge, MA: Harvard University Press, 2006), 99.

38 Albrecht, "Solastalgia," S95.

5 Disorientation in *The Maine Woods*

1 I reproduce Thoreau's prose in this passage as closely as possible, including strikeouts, misspellings, abbreviations, and unconventional punctuation. Thoreau, unpublished *Journal* transcript, 23:323, http://thoreau.library.ucsb.edu/writings_journals23.html. Accessed April 11, 2022.

2 Ibid., 23:323.

3 Ibid., 23:325.

4 Henry David Thoreau, *Letters to a Spiritual Seeker*, ed. Bradley P. Dean (New York: Norton, 2005), 158.

5 Thoreau interacted primarily with Penobscots, but the canoe routes of northern Maine were also used seasonally by other peoples, including the intertribal group known by Thoreau as "St. Francis Indians," or members of an allied group of Abenakis living in and around the village of Odanak, or St. Francis, on the St. Lawrence River in Quebec. "Wabanaki" refers to a group of confederated Algonquian nations in present-day northeastern New England, the Canadian Maritimes, and Quebec, including the Mi'kmaqs, Maliseets, Passamaquoddies, Penobscots, and several other tribes collectively known as the Western Abenakis, among others. This confederacy, dating to the sixteenth century, united these culturally similar tribes and villages in the wake of European contact. For a brief cultural introduction to the Penobscot, including the Wabanaki Confederacy, see Annette Kolodny, "A Summary History of the Penobscot Nation," in *The Life and Traditions of the Red Man*, ed. Joseph Nicolar and Annette Kolodny (Durham, NC: Duke University Press, 2007), 1–92. See also Frank G. Speck, "The Eastern Algonkian Wabanaki Confederacy," *American Anthropologist* 17 (1915), 492–508; and Frank G. Speck, *Penobscot Man: The Life History of a Forest Tribe in Maine*

(1940; reprint, with a new preface by David Sanger and additional photographs; Philadelphia: University of Pennsylvania Press, 1997), 16–24.

6 Thoreau records Polis's name for the stream, *Madunkehunk*, or Height of Land (*MW* 247). His companion, unnamed in the text, is Edward Hoar, the son of Concord lawyer and politician Samuel Hoar.

7 Thoreau, *Letters to a Spiritual Seeker*, 158.

8 Sophia Thoreau's role in editing Thoreau's posthumous publications has recently been brought to light by Kathy Fedorko, who proves that Ellery Channing and Ralph Waldo Emerson, long given credit by biographers and scholars, had no role in shaping *Excursions* (1863), *The Maine Woods* (1864), *Cape Cod* (1865), or *A Yankee in Canada, with Anti-Slavery and Reform Papers* (1866). Of *The Maine Woods*, Fedorko notes: "Textual evidence of Sophia's editing exists primarily in the essay 'The Allegash & East Branch,' where Sophia's handwriting appears frequently on the drafts: in the title, the expanded abbreviations, the rewriting in ink of Henry's pencil notations, the occasional word changes, and a transcription for the printer of all or part of Henry's last draft." Kathy Fedorko, " 'Henry's Brilliant Sister': The Pivotal Role of Sophia Thoreau in Her Brother's Posthumous Publications," *New England Quarterly* 89, no. 2 (2016), 243. Joseph Moldenhauer details Sophia's editing in *The Maine Woods*, as well. See Joseph J. Moldenhauer, "Editorial Appendix," in *MW* 449–56.

9 Laura Dassow Walls, " 'Some Star's Surface': Thoreau in the Maine Woods," in *Rediscovering the Maine Woods*, ed. John J. Kucich (Amherst: University of Massachusetts Press, 2019), 187.

10 Thoreau alludes to the 1846 trip in *Walden*, apparently finishing the draft shortly after leaving the woods. See Moldenhauer, "Editorial Appendix," in *MW* 357–9.

11 This element of surprise has been a consistent and important point in the extensive scholarship on "Ktaadn." David Robinson, for instance, notes that the essay "persistently records the undermining of the expectations Thoreau has brought to this excursion." David Robinson, *Natural Life: Thoreau's Worldly Transcendentalism* (Ithaca, NY: Cornell University Press, 2004), 135. See also Walls, *Thoreau*, 217–23.

12 Thoreau learned this early in the excursion: "When I asked McCauslin why more settlers did not come in, he answered, that one reason was, they could not buy the land, it belonged to individuals or companies who were afraid that their wild lands would be settled, and so incorporated into towns, and they be taxed for them; but to settling on the States' land there was no such hindrance. For his own part, he wanted no neighbors—he didn't wish to see any road by his house. Neighbors, even the best, were a trouble and expense, especially on the score of cattle and fences" (*MW* 24–5). This is an important lesson, and it remains a significant and understudied aspect of environmental thinking today, even as historic timber

industry landholdings in Maine and elsewhere have recently declined. See James S. Finley, "Pilgrimages and Working Forests: Envisioning the Commons in *The Maine Woods,*" in *Rediscovering the Maine Woods*, ed. John J. Kucich (Amherst: University of Massachusetts Press, 2019), 141–68.

13 *PJ* 2:315. This draft begins as a series of notes composed shortly after the trip, eventually growing into a fully formed narrative. Thoreau would continue to make changes as he lectured on the topic before publishing the piece in 1848 in the *Union Magazine of Literature and Art*. The final version was revised only lightly before appearing in *The Maine Woods*.

14 For an extended reading of this scene and a more detailed discussion of differences between this draft of "Ktaadn" and the essay's final form, see Jake McGinnis, "'A Smack of Wildness': Affect and Enchantment in Thoreau's First Taste of the Maine Woods," *Concord Saunterer* 26 (2018), 28–53.

15 Ktaadn, today usually Mount Katahdin, comes from the Penobscot, meaning "The Greatest Mountain." In this chapter, I defer to Thoreau's spelling and omit as redundant "Mount" from the name.

16 "disorient, v." *OED Online*, December 2021. Oxford University Press.

17 See, for example, *The Norton Book of Nature Writing* (New York: W.W. Norton, 2002), and *Literature and Nature* (Upper Saddle River, NJ: Prentice Hall, 1996); Greg Garrard, *Ecocriticism: The New Critical Idiom* (New York: Routledge, 2004); Roderick Nash, *Wilderness and the American Mind* (New Haven, CT: Yale University Press, 1967); and William Cronon, "The Trouble with Wilderness; or, Getting Back to the Wrong Nature," in *Uncommon Ground: Rethinking the Human Place in Nature*, ed. William Cronon (New York: W.W. Norton, 1996), 69–90. See also Bradley P. Dean, "Natural History, Romanticism, and Thoreau," in *American Wilderness: A New History*, ed. Michael Lewis (New York: Oxford University Press, 2007), 73–90.

18 On Thoreau and the sublime, see Ronald Wesley Hoag's landmark essay, "The Mark on the Wilderness: Thoreau's Contact with Ktaadn," *Texas Studies in Literature and Language* 24, no. 1 (1982), 23–46. See also Edmund Burke, *Philosophical Enquiry into the Origin of Our Ideas of the Sublime and Beautiful*, ed. Adam Phillips (Oxford: Oxford University Press, 1998; originally published 1756).

19 Thoreau may have made a map of his route on the mountain, but if so, it has been lost. See William W. Geller, "'A Thoreau Drawn Diagram of Ktaadn': If It Exists, Where Is It?" in *Maine History Documents* 298 (Bangor: University of Maine Commons, 2019).

20 Ronald Wesley Hoag, "On Thoreau's Trail in the Maine Woods: A *Photographic Journey* Worth Taking, Critics Howling in the Wilderness," *Thoreau Society Bulletin*

285 (Spring 2014), 3. See also Thoreau, "Sublimity," in *Early Essays and Miscellanies*, ed. Joseph J. Moldenhauer (Princeton, NJ: Princeton University Press, 1975), 93–8.

21 Christopher Hitt, "Toward an Ecological Sublime," *New Literary History* 30, no. 3 (1999), 617.

22 See McGinnis, "A Smack of Wildness," 34.

23 Hoag, "The Mark on the Wilderness," 24.

24 Melissa Sexton, "Sublime Matter: Materiality and Language in Thoreau's Ktaadn," in *Rediscovering the Maine Woods: Thoreau's Legacy in an Unsettled Land*, ed. John J. Kucich (Amherst: University of Massachusetts Press, 2019), 107. See also James S. Finley, "Matter and Objects in the Maine Woods," *Concord Saunterer* 22 (2014), 22–9; and David Robinson, "Thoreau's 'Ktaadn' and the Quest for Experience," in *Emersonian Circles*, ed. Robert Burkholder and Wesley T. Mott (Rochester, NY: University of Rochester Press, 1997), 207–24.

25 Sexton, "Sublime Matter," 109.

26 These convergences took many forms, from Thoreau's interest in earthly dynamism to similar passages and allusions that exist within each text. For example, explicit references and more subtle gesture to Melville's *Typee* occur in each of these texts, originating in an early draft of "Ktaadn" before making their way into both *A Week* and *Walden*. See McGinnis, "Peeping at Melville's *Typee*: Thoreau's Contact with 'Dusky and Unexplored Nature,'" *ESQ* 66, no. 2 (2020), 562–89.

27 On Thoreau's attitude toward hunting, see Thomas L. Altherr, "'Chaplain to the Hunters': Henry David Thoreau's Ambivalence Toward Hunting," *American Literature* 56, no. 3 (1984), 345–61; Michael Stoneham, "Thoreau's Hunting Ethos: Teaching Right Natural Relations in Modern America," *Thoreau Society Bulletin* 282 (2013), 8–10; and Jericho Williams, "Trigger Warning: Thoreau, 'Chesuncook,' and the Complications of Hunting," *Thoreau Society Bulletin*, no. 286 (2014), 1–3.

28 On Lowell's changes to the manuscript in its 1858 appearance in *The Atlantic*, see Moldenhauer, "Editorial Appendix," in *MW* 360–2.

29 Robert F. Sayre, *Thoreau and the American Indians* (Princeton, NJ: Princeton University Press, 1977), 171.

30 *Eliot's Indian Bible* (*Mamusse Wunneetupanatamwe Up-Biblum God*), a translation of the New Testament into the Massachusetts language, was published by John Eliot in 1661, the first translation of its kind in the Americas and the first Bible to be printed in British North America. It was expanded in 1663 to include the New Testament. Paugus, chief of the Pequawket, was killed in 1725 in the Battle of Pequawket, the last major engagement of the Dummer's War, an eighteenth-century conflict between Wabanaki and English settlers over colonial expansion.

31 See Robert Sattelmeyer, *Thoreau's Reading: A Study in Intellectual History with Bibliographical Catalog* (Princeton, NJ: Princeton University Press, 1988), 99–110.

32 The Maine historian, writer, folklorist, and ornithologist Fannie Hardy Eckstorm identifies one of these other men as Sebattis Dana of Oldtown, and "originally a Maliseet from Quoddy." Fannie Hardy Eckstorm, "Notes on Thoreau's *Maine Woods*," *Thoreau Society Bulletin* 51 (Spring 1955), 1.

33 See Moldenhauer, "Editorial Appendix," 363–4.

34 Sayre, *Thoreau and the American Indians*, 172.

35 Ibid., 155, 158. Linda Frost also suggests that the Maine essays together demonstrate an evolving appreciation for Indigenous peoples. Linda Frost, "The Red Face of Man, the Penobscot Indian and Conflict of Interest in Thoreau's *The Maine Woods*," *ESQ* 39, no. 1 (1993), 21–47.

36 Joshua Bellin, "In the Company of Savagists: Thoreau's Indian Books and Antebellum Ethnology," *Concord Saunterer* 16 (2008), 2. These commonplace books, often referred to as the "Indian Books," have obtained an almost mythic status among some Thoreau specialists, representing "nearly 3000 manuscript pages of extracts from historical, ethnological, missionary, and other writings on the Indians interspersed with occasional authorial commentary" (2). Although largely unpublished, these commonplace books have attracted a great deal of scholarly attention. See, for example, Albert Keiser, *The Indian in American Literature* (Oxford: Oxford University Press, 1933); Lawrence Willson, "Thoreau: Student of Anthropology," *American Anthropologist* 61, no. 2 (April 1959), 279–89; Sayre, "An Indian Book?" in *Thoreau and the American Indians*, 101–22; Richard Fleck, *The Indians of Thoreau: Selections from the Indian Notebooks* (Albuquerque: Hummingbird, 1974); and Suzanne Dvorak Rose, "Tracking the Moccasin Print: A Descriptive Index to Henry David Thoreau's Indian Notebooks and a Study of the Relationship of the Indian Notebooks to Mythmaking in Walden" (PhD diss., University of Oklahoma, 1994).

37 John J. Kucich, *Unsettling Thoreau: Native Americans, Settler Colonialism, and the Power of Place* (Amherst: University of Massachusetts Press, 2024), 141.

38 Richard Schneider, *Civilizing Thoreau: Human Ecology and the Emerging Social Sciences in the Major Works* (Rochester, NY: Camden House, 2016), 37.

39 Michael Stoneham, "Remeasuring Thoreau: *The Maine Woods* and Thoreau's Evolving Appreciation of the Racial Other," *Concord Saunterer* 27 (2019), 74.

40 Kucich, *Unsettling Thoreau*, 141.

41 Kucich lists these attributes: "a sense of on-going creation that counters Western conceptions of teleology; a conception of a natural world that is alive, full of spiritual power and endowed with personhood; and a firm sense of the reciprocal

relationship between humans and the environment that later European-American writers would term a 'land ethic.'" Kucich, *Unsettling Thoreau*, 163.

42 This "mighty Indian hunter," Gluskabe or Klose-kur-beh, is part of a story rendered physically present on the landscape, visible in the topography of the Maine woods, and particularly at Kineo, above Moosehead Lake. See Joseph Nicolar, *The Life and Traditions of the Red Man*, ed. Annette Kolodny (Durham, NC: Duke University Press, 2007, originally published 1893). See also the bilingual map and gazetteer, "Iyoka Eli-Wihtamakw Kətahkinawal/This is How We Name Our Lands" (Indian Island, ME: Penobscot Nation Cultural and Historic Preservation Department, 2015).

43 For the extensive revisions that Thoreau made in his original account of this moment, see Thoreau, unpublished *Journal* transcript, 23:322 cf.

44 Ibid., 23:327.

45 Ibid., 23:331.

46 Ibid., 23:331.

47 Sayre, *Thoreau and the American Indians*, 181.

48 Ibid., 180.

49 Kucich, *Unsettling Thoreau*, 165.

50 Ibid., 165.

51 Thoreau, *Letters to a Spiritual Seeker*, 153–4.

Coda: Excursions in the Anthropocene

1 Barry Lopez, *Horizon* (New York: Knopf, 2019), 108.

2 Ibid., 109.

3 Ibid., 23

4 Ibid.

5 Edward Abbey, *Desert Solitaire* (Tucson: University of Arizona Press, 1988), 39–55.

6 Ibid., 23.

7 Ibid., 109.

8 Ibid.

9 Ibid., 125.

10 Ibid., 126.

11 Ibid.

12 Ibid.

13 Ibid., 127.

14 Ibid.

15 Ibid., 26.

16 Michiko Kakutani, "Books of the Times," *New York Times*, February 12, 1986, C21.

17 Ibid.

18 Barry Lopez, "An Era of Emergencies Is Upon Us and We Cannot Look Away," in *Embrace Fearlessly the Burning World: Essays* (New York: Random House, 2022), 21–6.

19 Lopez, *Horizon*, 129.

Bibliography

Abbey, Edward. *Desert Solitaire*. Tucson: University of Arizona Press, 1988.

Abram, David. *Spell of the Sensuous: Perception and Language in a More-Than-Human-World*. New York: Vintage Books, 1996.

Abrams, Robert E. *Landscape and Ideology in American Renaissance Literature: Topographies of Skepticism*. Cambridge: Cambridge University Press, 2004.

Abrams, Robert E. "Thoreau Catching Cold: 'A Yankee in Canada.'" *Emerson Society Quarterly* 25, no. 97 (1979): 224–34.

Adams, Stephen. "'That Tidiness We Always Look for in Woman': Fuller's *Summer on the Lakes* and Romantic Aesthetics." *Studies in the American Renaissance* (1987): 247–64.

Ahmed, Sara. "Happy Objects." In *Affect Theory Reader*, edited by Gregory Seigworth and Melissa Gregg, 29–51. Durham, NC: Duke University Press, 2010.

Albrecht, Glenn, Gina-Maree Sartore, Linda Connor, Nick Higginbotham, Sonia Freeman, Brian Kelly, Helen Stain, Anne Tonna, and Georgia Pollard, "Solastalgia: The Distress Caused by Environmental Change." *Australian and New Zealand Journal of Psychiatry* 15, no. S1 (2007), S95–8.

Allen, Margaret V. *The Achievement of Margaret Fuller*. University Park: Pennsylvania State University Press, 1979.

Allen, Margaret V. "The Political and Social Criticism of Margaret Fuller." *South Atlantic Quarterly* 72, no. 4 (1973): 560–73.

Altherr, Thomas L. "'Chaplain to the Hunters': Henry David Thoreau's Ambivalence Toward Hunting." *American Literature* 56, no. 3 (1984): 345–61.

Antelyes, Peter. *Tales of Adventurous Enterprise: Washington Irving and the Poetics of Western Expansion*. New York: Columbia University Press, 1990.

Arsić, Branka. *Bird Relics: Grief and Vitalism in Thoreau*. Cambridge, MA: Harvard University Press, 2016.

Avallone, Charlene. "The Red Roots of White Feminism in Margaret Fuller's Writings." In *Doing Feminism: Teaching and Research in the Academy*, edited by Mary Anderson, Lisa Fine, Kathleen Geissler, and Joyce R. Landenson, 135–65. East Lansing: Michigan University Press, 1997.

Bahnson, Fred. "The World We Still Have: Barry Lopez on Restoring Lost Intimacy With Nature." *The Sun*, December 2019. www.thesunmagazine.org/issues/528/the-world-we-still-have. Accessed July 15, 2020.

Baker, Anne. "'A Commanding View': Vision and the Problem of Nationality in Fuller's *Summer on the Lakes*." *ESQ: A Journal of the American Renaissance* 44, nos. 1–2 (1998): 61–78.

Baraga, Frederic. *A Dictionary of the Otchipwe Language*. 2 vols. Montreal: Beauchemin & Valois, 1878.

Bartol, C. A. *Pictures of Europe, Framed in Ideas*. Boston: Crosby, Nichols, 1856.

Basile, Joseph Lawrence. "Thoreau's Uncommon Cold: 'A Yankee in Canada' Revisited." *Concord Saunterer* 13, no. 1 (1978): 7–11.

Bean, Judith Matteson. "'A *Presence* Among Us': Fuller's Place in Nineteenth-Century Oral Culture." *ESQ: A Journal of the American Renaissance* 44, nos. 1–2 (1998): 79–123.

Bellin, Joshua. "In the Company of Savagists: Thoreau's Indian Books and Antebellum Ethnology." *Concord Saunterer* 16 (2008): 1–32.

Berkhofer, Robert F., Jr. *The White Man's Indian: Images of the American Indian from Columbus to the Present*. New York: Knopf, 1978.

Berlant, Lauren Gail. *Cruel Optimism*. Durham, NC: Duke University Press, 2011.

Bevis, William. "Native American Novels: Homing In." In *Recovering the Word: Essays on Native American Literature*, edited by Brian Swann and Arnold Krupat, 580–620. Berkeley: University of California Press, 1987.

Blackbird, Andrew. *History of the Ottawa and Chippewa Indians of Michigan*. Ypsilanti, MI: Ypsilanti Auxiliary of the Woman's National Indian Association, 1887.

Boudreau, Gordon V. *The Roots of Walden and the Tree of Life*. Nashville, TN: Vanderbilt University Press, 1990.

Breitwieser, Mitchell Robert. "Thoreau and the Wrecks on Cape Cod." In *National Melancholy: Mourning and Opportunity in Classic American Literature*, 144–59. Stanford: Stanford University Press, 2007.

Bridgman, Richard. *Dark Thoreau*. Lincoln: University of Nebraska Press, 1982.

Brodhead, Richard H. *Cultures of Letters: Scenes of Reading and Writing in Nineteenth-Century America*. Chicago: University of Chicago Press, 1993.

Brown, William Wells. *The American Fugitive in Europe Sketches of Places and People Abroad*. Boston: John P. Jewett, 1855.

Brown, William Wells. *Three Years in Europe: Or, Places I Have Seen and People I Have Met*. London: C. Gilpin, 1852.

Brownson, Orestes. "Miss Fuller and Reformers." In *Critical Essays on Margaret Fuller*, edited by Joel Myerson, 19–25. Boston: G.K. Hall, 1980.

Brownson, Orestes. Review of *Summer on the Lakes*. In *Critical Essays on Margaret Fuller*, edited by Joel Myerson, 5–6. Boston: G.K. Hall, 1980.

Buell, Lawrence. *The Environmental Imagination: Thoreau, Nature Writing, and the Formation of American Culture*. Cambridge, MA: Harvard University Press, 1995.

Buell, Lawrence. *The Future of Environmental Criticism: Environmental Crisis and Literary Imagination*. Malden, MA: Blackwell, 2005.

Buell, Lawrence. "Henry Thoreau Enters the American Canon." In *New Essays on Walden*, edited by Robert F Sayre, 23–55. Cambridge: Cambridge University Press, 1992.

Buell, Lawrence. *Literary Transcendentalism: Style and Vision in the American Renaissance*. Ithaca, NY: Cornell University Press, 1973.

Buell, Lawrence. *The Pilgrim's Progress from This World to That Which Is to Come ….* London: Nath. Ponder, 1678.

Bunyan, John. *Grace Abounding to the Chief of Sinners, or, A Brief and Faithful Relation of the Exceeding Mercy of God in Christ, to His Poor Servant John Bvnyan …* London: George Larkin, 1666.

Burke, Edmund. *Philosophical Enquiry into the Origin of Our Ideas of the Sublime and Beautiful*. Edited by Adam Phillips. Oxford: Oxford University Press, 1998.

Burns, Mark K. "Ineffectual Chase: Indians, Prairies, Buffalo, and the Quest for the Authentic West in Washington Irving's *A Tour on the Prairies*." *Western American Literature* 42, no. 1 (Spring 2007): 54–79.

Cameron, Kenneth Walter, ed. *Thoreau's Canadian Notebook, and Record of Surveys: Selected Chapters from Transcendental Climate*. Hartford, CT: Transcendental Books, 1967.

Cameron, Kenneth Walter. *Transcendental Climate; New Resources for the Study of Emerson, Thoreau and Their Contemporaries*. Hartford, CT: Transcendental Books, 1963.

Capper, Charles. *Margaret Fuller: An American Romantic Life*. 2 vols. New York: Oxford University Press, 1992.

Capper, Charles. "Margaret Fuller as Cultural Reformer: The Conversations in Boston." *American Quarterly* 39, no. 4 (1987): 509–28.

Carlyle, Thomas. *Chartism*. Boston: Charles C. Little and James Brown, 1840.

Carlyle, Thomas. *Past and Present*. Boston: Charles C. Little and James Brown, 1843.

Carter, Erica, James Donald, and Judith Squires, eds. *Space and Place: Theories of Identity and Location*. London: Lawrence and Wishart, 1993.

Case, Kristen. "Knowing as Neighboring: Approaching Thoreau's Kalendar." *J19: The Journal of Nineteenth-Century Americanists* 2, no. 1 (2014): 107–29.

Catlin, George. *Letters and Notes on the Manners, Customs, and Conditions of the North American Indians: Written During Eight Years' Travel (1832–1839) Amongst the Wildest Tribes of Indians in North America*. 2 vols. Edited by Marjorie Halpin. New York: Dover, 1973.

"Ceded Territory Boundary" (Treaties of 1836, 1837, 1842 and 1854) [Map]. Odanah, WI: Great Lakes Indian Fish and Wildlife Commission, January 2015. https://data.glifwc.org/ceded/. Accessed April 7, 2022.

Chakrabarty, Dipesh. "Climate and Capital: On Conjoined Histories." *Critical Inquiry* 41 (2014): 1–23.

Chakrabarty, Dipesh. "The Climate of History: Four Theses." *Critical Inquiry* 35 (2009): 197–222.

Chakrabarty, Dipesh. "Humanities in the Anthropocene: The Crisis of the Enduring Kantian Fable." *New Literary History* 47, nos. 2–3 (2016): 377–97.

Christie, John Aldrich. *Thoreau as World Traveler*. New York: Columbia University Press, 1965.

Clapper, Ronald Earl. "The Development of Walden: A Genetic Text." PhD diss., University of California, 1967.

Clark, Timothy. *The Cambridge Introduction to Literature and Environment*. Cambridge: Cambridge University Press, 2011.

Clarke, James Freeman. *Eleven Weeks in Europe; and What May Be Seen in That Time*. Boston: Ticknor, Reed & Fields, 1852.

Clarke, James Freeman. Review of *Summer on the Lakes*. In *Critical Essays on Margaret Fuller*, edited by Joel Myerson. Boston: G.K. Hall, 1980.

Cohen, Jeffrey Jerome, and Lowell Duckert, eds. *Veer Ecology: A Companion for Environmental Thinking*. Minneapolis: University of Minnesota Press, 2017.

Cohen, Jeffrey Jerome, and Lowell Duckert. "Welcome to the Whirled." In *Veer Ecology: A Companion for Environmental Thinking*, edited by Jeffrey Jerome Cohen and Lowell Duckert, 1–15. Minneapolis: University of Minnesota Press, 2017.

Cole, Phyllis. "The Nineteenth-Century Women's Rights Movement and the Canonization of Margaret Fuller." *ESQ: A Journal of the American Renaissance* 44, nos. 1–2 (1998): 1–34.

Conn, Steven. *History's Shadow: Native Americans and Historical Consciousness in the Nineteenth Century*. Chicago: University of Chicago Press, 2004.

Copway, George. *Life, Letters, and Speeches of George Copway (Kahgegagahbowh)*. Edited by A. LaVonne Brown and Donald B. Smith. Lincoln: University of Nebraska Press, 1997.

Copway, George. *Organization of a New Indian Territory East of the Missouri River Arguments and Reasons Submitted to the Honorable the Members of the Senate and House of Representatives of the 31st Congress of the United States*. New York: S.W. Benedict, 1850.

Copway, George. *Recollections of a Forest Life: Life, History, and Travels of Kah-ge-ga-gah-bowh*. Edited by Deanna Reader. Waterloo, ON: Wilfrid Laurier University Press, 2022.

Copway, George. *Running Sketches of Men and Places in England, France, Germany, Belgium, and Scotland*. New York: J. C. Riker, 1851.

Copway, George. *Traditional History and Characteristic Sketches of the Ojibway Nation*. London: C. Gilpin, 1850.

Cronon, William. *Changes in the Land: Indians, Colonists, and the Ecology of New England*. New York: Hill and Wang, 1983.

Cronon, William. "The Trouble with Wilderness; or, Getting Back to the Wrong Nature." In *Uncommon Ground: Rethinking the Human Place in Nature*, edited by William Cronon, 66–90. New York: W. W. Norton, 1996.

Curtis, George William. *The Howadji in Syria*. New York: Harper & Brothers, 1852.

Curtis, George William. *Lotus-Eating: A Summer Book*. New York: Harper & Brothers, 1852.

Curtis, George William. *Nile Notes of a Howadji*. New York: Harper & Brothers, 1851.

Dall, Caroline W. Healey. *Margaret and Her Friends*. Boston: Roberts Brothers, 1895.

Dean, Bradley P. "Natural History, Romanticism, and Thoreau." In *American Wilderness: A New History*, edited by Michael Lewis, 73–90. New York: Oxford University Press, 2007.

Deloria, Philip. *Playing Indian*. New Haven, CT: Yale University Press, 1998.

Dimick, Sarah. "Disordered Environmental Time: Phenology, Climate Change, and Seasonal Form in the Work of Henry David Thoreau and Aldo Leopold." *ISLE: Interdisciplinary Studies in Literature and Environment* 25, no. 4 (2018): 700–21.

Dippie, Brian W. *The Vanishing American: White Attitudes and the U.S. Indian Policy*. Berkeley: University of California Press, 1982.

"disorient, v." *OED Online*. December 2021. Oxford University Press. www-oed-com.proxy.library.nd.edu/view/Entry/54878?redirectedFrom=disorient#eid.

Donahue, Brian. *The Great Meadow: Farmers and the Land in Colonial Concord*. New Haven, CT: Yale University Press, 2004.

Doyle, Kathleen Coyne. "Thoreau's Flute, or, 'the Moose, the Pine Tree, & the Indian.'" *Concord Saunterer* 22 (2014): 99–111.

Drake, Joseph Rodman. *The Culprit Fay: and Other Poems*. New York: G. Dearborn, 1835.

Eckstorm, Fannie Hardy. "Notes on Thoreau's *Maine Woods*." *Thoreau Society Bulletin* 51 (Spring 1955): 1, 3.

Elbert, Monika. "Haunting Transcendentalist Landscapes: EcoGothic Politics in Margaret Fuller's *Summer on the Lakes*." *Text Matters* 6, no. 6 (2016): 53–73.

Ellis, Cristin. *Antebellum Posthuman: Race and Materiality in the Mid-Nineteenth Century*. New York: Fordham University Press, 2018.

Ellsworth, Henry Leavitt. *Washington Irving on the Prairie; or, A Narrative of the Tour of the Southwest in the Year 1832*. Edited by Stanley T. Williams and Barbara D. Simison. New York: American Book Company, 1937.

Emerson, Ralph Waldo. *English Traits*. Boston: Phillips, Sampson, 1856.

English, Karen A. "'Genuine Transcripts of Private Experience': Margaret Fuller and Translation." *American Transcendental Quarterly* 15, no. 2 (2001): 131–8.

"excursion, n." *OED Online*. December 2020. Oxford University Press. www-oed-com.proxy.library.nd.edu/view/Entry/65949?rskey=vdJtB3&result=1.

Fedorko, Kathy. "'Henry's Brilliant Sister': The Pivotal Role of Sophia Thoreau in Her Brother's Posthumous Publications." *New England Quarterly* 89, no. 2 (2016): 222–56.

Finch, Robert, and John Elder, eds. *The Norton Book of Nature Writing*. New York: W. W. Norton, 1990.

Finley, James S. "Matter and Objects in the Maine Woods." *Concord Saunterer* 22 (2014): 22–9.

Finley, James S. "Pilgrimages and Working Forests: Envisioning the Commons in *The Maine Woods*." In *Rediscovering the Maine Woods*, edited by John J. Kucich, 141–68. Amherst: University of Massachusetts Press, 2019.

Fish, Cheryl J. *Black and White Women's Travel Narratives*. Gainesville: University of Florida Press, 2004.

Fleck, Richard. *The Indians of Thoreau: Selections from the Indian Notebooks*. Albuquerque, NM: Hummingbird, 1974.

Foster, Edward Halsey. *The Civilized Wilderness: Backgrounds to American Romantic Literature, 1817–1860*. New York: Free Press, 1975.

Franklin, Wayne. *Discoverers, Explorers, Settlers: The Diligent Writers of Early America*. Chicago: University of Chicago Press, 1979.

Frost, Linda. "The Red Face of Man, the Penobscot Indian and Conflict of Interest in Thoreau's *The Maine Woods*." *ESQ* 39, no. 1 (1993): 21–47.

Fuller, Margaret. *At Home and Abroad, or, Things and Thoughts in America and Europe*. Edited by Arthur B. Fuller. Boston: Crosby, Nichols, 1856.

Fuller, Margaret. *The Letters of Margaret Fuller*. 6 vols. Edited by Robert N. Hudspeth. Ithaca, NY: Cornell University Press, 1983–95.

Fuller, Margaret. *Summer on the Lakes, in 1843*. Edited by Susan Belasco Smith. Urbana: University of Illinois Press, 1991.

Garrard, Greg. *Ecocriticism: The New Critical Idiom*. New York: Routledge, 2004.

Geller, William W. "'A Thoreau Drawn Diagram of Ktaadn': If It Exists, Where Is It?" In *Maine History Documents* 298. Bangor: University of Maine Commons, 2019. https://digitalcommons.library.umaine.edu/mainehistory/298. Accessed July 7, 2024.

Giddens, Anthony, and Christopher Pierson. *Conversations with Anthony Giddens: Making Sense of Modernity*. Stanford, CA: Stanford University Press, 1999.

Giles, Paul. "Transnational Thoreau: Time, Space, and Relativity." In *Thoreau at 200: Essays and Reassessments*, edited by Kristen Case and K. P. Van Anglen, 138–53. Cambridge: Cambridge University Press, 2016.

Gould, Rebecca Kneale. "Thoreau, Race and Environmental Justice: Deepening the Conversation." *Concord Saunterer* 25 (2017): 171–83.

Gura, Philip F. *American Transcendentalism: A History*. New York: Hill and Wang, 2007.

Gustafson, Sandra. "Choosing a Medium: Margaret Fuller and the Forms of Sentiment." *American Quarterly* 47, no. 1 (March 1995): 34–65.

Harding, Walter. *The Days of Henry David Thoreau: A Biography*. Princeton, NJ: Princeton University Press, 1992.
Harding, Walter. *A Thoreau Handbook*. New York: New York University Press, 1959.
Haronian, Mary-Jo. "Margaret Fuller's Visions." *ESQ: A Journal of the American Renaissance* 44, nos. 1–2 (1998): 35–60.
Hessler, John, ed. "From Ortelius to Champlain: The Lost Maps of Henry David Thoreau." *Concord Saunterer* 18 (2010): 1–26.
Hessler, John. *Henry David Thoreau on Cape Cod*. Delray Beach, FL: Levenger Press, 2010.
Hillard, George Stillman. *Six Months in Italy*. 2 vols. Boston: Ticknor, Reed, and Fields, 1853.
Hitt, Christopher. "Toward an Ecological Sublime." *New Literary History* 30, no. 3 (1999): 603–23.
Hoag, Ronald Wesley. "The Mark on the Wilderness: Thoreau's Contact with Ktaadn." *Texas Studies in Literature and Language* 24, no. 1 (1982): 23–46.
Hoag, Ronald Wesley. "On Thoreau's Trail in the Maine Woods: A *Photographic Journey* Worth Taking, Critics Howling in the Wilderness." *Thoreau Society Bulletin* 285 (Spring 2014): 1–5.
Huth, Hans. *Nature and the American: Three Centuries of Changing Attitudes*. Lincoln: University of Nebraska Press, 1957.
Irving, Washington. "The Adventures of Captain Bonneville." In *Three Western Narratives*, edited by James P. Ronda, 617–956. New York: Library of America, 2004.
Irving, Washington. "The Alhambra." In *Bracebridge Hall, Tales of a Traveller, The Alhambra*, edited by Andrew Myers, 719–1050. New York: Library of America, 1991.
Irving, Washington. "Astoria." In *Three Western Narratives*, edited by James P. Ronda, 163–616. New York: Library of America, 2004.
Irving, Washington. "Bracebridge Hall, or The Humorists, A Medley." In *Bracebridge Hall, Tales of a Traveller, The Alhambra*, edited by Andrew Myers, 1–378. New York: Library of America, 1991.
Irving, Washington. *A Chronicle of the Conquest of Granada*. London: J. Murray, 1829.
Irving, Washington. *A History of the Life and Voyages of Christopher Columbus*. New York: G. & C. Carvill, 1828.
Irving, Washington. "The Sketchbook of Geoffrey Crayon, Gent." In *History, Tales, and Sketches*, edited by James Tuttleton, 732–1091. New York: Library of America, 1983.
Irving, Washington. "Tales of a Traveller." In *Bracebridge Hall, Tales of a Traveller, The Alhambra*, edited by Andrew Myers, 379–718. New York: Library of America, 1991.
Irving, Washington. "A Tour on the Prairies." In *Three Western Narratives*, edited by James P. Ronda, 1–162. New York: Library of America, 2004.
Irving, Washington. *Voyages and Discoveries of the Companions of Columbus*. Philadelphia, PA: Carey and Lea, 1831.

Irving, Washington. *The Western Journals of Washington Irving*. Edited by John Francis McDermott. Norman: University of Oklahoma Press, 1944.

"Iyoka Eli-Wihtamakw Kətahkinawal/This Is How We Name Our Lands." Indian Island, ME: Penobscot Nation Cultural and Historic Preservation Department, 2015.

Jameson, Anna. *Winter Studies and Summer Rambles in Canada*. New York: Wiley and Putnam, 1839.

Johnson, Linck. "Into History: Thoreau's Earliest 'Indian Book' and the His First Trip to Cape Cod." *Emerson Society Quarterly* 107 (1982): 75–88.

Johnson, Linck. *Thoreau's Complex Weave: The Writing of A Week on the Concord and Merrimack Rivers*. Charlottesville: University Press of Virginia, 1986.

Johnson, Rochelle L. "'This Enchantment Is No Delusion': Henry David Thoreau, the New Materialisms, and Ineffable Materiality." *ISLE: Interdisciplinary Studies in Literature and Environment* 21, no. 3 (2014): 606–35.

Johnson, Samuel. *Johnson's English Dictionary*. Philadelphia, PA: Kimber & Sharpless, 1841.

Jones, Jamie L. "Long Looking, Without Seeing: Thoreau's *Cape Cod* and Global Capitalism at Sea." *The Dial: A Journal of the Ralph Waldo Emerson Society* 1, no. 1 (2024): 42–65.

Jones, Peter. *History of the Ojebway Indians: With Especial Reference to Their Conversion to Christianity*. London: A. W Bennett, 1861.

Judd, Richard William. *Second Nature: an Environmental History of New England*. Amherst: University of Massachusetts Press, 2014.

Kakutani, Michiko. "Books of the Times" [Review of *Arctic Dreams*]. *New York Times*, February 12, 1986, C21.

Kappler, Charles J., ed. *Indian Affairs: Laws and Treaties*, 7 vols. Washington, DC: Government Printing Office, 1904.

Keegan, Bridget, and James C. McKusick, eds. *Literature and Nature: Four Centuries of Nature Writing*. Upper Saddle River, NJ: Prentice Hall, 2001.

Keiser, Albert. *The Indian in American Literature*. Oxford: Oxford University Press, 1933.

Kime, Wayne R. "The Author as Professional: Washington Irving's 'Rambling Anecdotes' of the West." In *Critical Essays on Washington Irving*, edited by Ralph M. Aderman, 237–53. Boston: G.K. Hall, 1990.

Kolodny, Annette. *The Land Before Her: Fantasy and Experience of the American Frontiers, 1630–1860*. Chapel Hill: University of North Carolina Press, 1984.

Kolodny, Annette. "A Summary History of the Penobscot Nation." In *The Life and Traditions of the Red Man*, edited by Joseph Nicolar and Annette Kolodny, 1–92. Durham, NC: Duke University Press, 2007.

Konkle, Maureen. *Writing Indian Nations: Native Intellectuals and the Politics of Historiography, 1827–1863*. Chapel Hill: University of North Carolina Press, 2004.

Kucich, John, J. "An Imperfect Indian Wisdom: Thoreau, Ecocultural Contact, and the Spirit of Place." In *Thoreau Beyond Borders: New International Essays on America's Most Famous Nature Writer*, edited by Francois Specq, Laura Dassow Walls, and Julien Negre, 9–24. Amherst: University of Massachusetts Press, 2020.

Kucich, John, J. "Lost in the Maine Woods: Henry David Thoreau, Joseph Nicolar, and the Penobscot World." *Concord Saunterer* 19/20 (2011–12): 22–52.

Kucich, John J., ed. *Rediscovering the Maine Woods: Thoreau's Legacy in an Unsettled Land*. Amherst: University of Massachusetts Press, 2019.

Lacroix, Patrick. "Finding Thoreau in French Canada: The Ideological Legacy of the American Revolution." *American Review of Canadian Studies* 47, no. 3 (2017): 266–79.

Larouche, Leon-Maurice. "For the Celebration of the 150th Anniversary of 'A Yankee in Canada': The Viewpoint of a Thoreauvian in Quebec." *Thoreau Society Bulletin* 230 (2000): 1–2.

Latrobe, Charles Joseph. *The Rambler in North America: 1832–33*. 2 vols. London: R. B. Seeley and W. Burnside, 1835.

Lemire, Elise. *Black Walden: Slavery and Its Aftermath in Concord, Massachusetts*. Philadelphia: University of Pennsylvania Press, 2009.

Lenckos, Elisabeth. " 'Stimulus and Cheer': Margaret Fuller's 'Translations,' from Eckermann's Conversations with Goethe to Bettina von Arnim's Guenderode." In *Translators, Interpreters, Mediators: Women Writers 1700–1900*, edited by Gillian E. Dow, 191–207. Bern: Peter Lang, 2007.

Lilley, James D. *Common Things: Romance and the Aesthetics of Belonging in Atlantic Modernity*. New York: Fordham University Press, 2014.

Lopez, Barry. "An Era of Emergencies Is Upon Us and We Cannot Look Away." In *Embrace Fearlessly the Burning World: Essays*, 21–6. New York: Random House, 2022.

Lopez, Barry. *Horizon*. New York: Knopf, 2019.

Lynch, Tom, Cheryll Glotfelty, and Karla Armbruster, eds. *The Bioregional Imagination: Literature, Ecology, and Place*. Athens: University of Georgia Press, 2012.

Lynch, Tom, Susan Naramore Maher, Drucilla Wall, and O. Alan Weltzien. "An Alignment of Stones." In *The Bioregional Imagination: Literature, Ecology, and Place*, edited by Tom Lynch, Susan Naramore Maher, Drucilla Wall, and O. Alan Weltzien, xi–xxvi. Lincoln: University of Nebraska Press, 2017.

Lyons, Scott Richard. "Migrations to Modernity: The Many Voices of George Copway's *Running Sketches of Men and Places, in England, France, Germany, Belgium, and Scotland*." In *The World, the Text, and the Indian: Global Dimensions of Native American Literature*, edited by Scott Richard Lyons, 143–82. New York: SUNY Press, 2017.

Lyons, Scott Richard. *X-Marks: Native Signatures of Assent*. Minneapolis: University of Minnesota Press, 2010.

Maddox, Lucy. *Removals: Nineteenth-Century American Literature and the Politics of Indian Affairs*. Oxford: Oxford University Press, 1991.

Mark, Jason. "In Defense of Henry David Thoreau: New Literary Revisionism Might Make the Seer of Walden Pond More Relevant to Readers Today." *Sierra*, October 27, 2015. www.sierraclub.org/sierra/2015-5-september-october/green-life/defense-henry-david-thoreau. Accessed July 15, 2025.

Marsh, George Perkins. *Man and Nature*. Edited by David Lowenthal. Seattle: University of Washington Press, 2003.

Marshall, Megan. *Margaret Fuller: A New American Life*. Boston: Houghton Mifflin Harcourt, 2013.

McGinnis, Jake. "Peeping at Melville's *Typee*: Thoreau's Contact with 'Dusky and Unexplored Nature.'" *ESQ: A Journal of Nineteenth-Century Literature and Culture* 66, no. 2 (2020): 562–89.

McGinnis, Jake. "'A Smack of Wildness': Affect and Enchantment in Thoreau's First Taste of the Maine Woods." *Concord Saunterer* 26 (2018): 28–53.

McKenney, Thomas Loraine. *Sketches of a Tour to the Lakes, of the Character and Customs of the Chippeway Indians, and of Incidents Connected with the Treaty of Fond Du Lac*. Baltimore, MD: Fielding Lucas, Jun'r., 1827.

McNeil, J.R. "Introductory Remarks: The Anthropocene and the Eighteenth Century." *Eighteenth-Century Studies* 49, no. 2 (2016): 117–28.

Melville, Herman. *Typee: A Peep at Polynesian Life*. New York: Penguin Books, 1996.

Miller, Perry. *Margaret Fuller: American Romantic*. New York: Doubleday, 1963.

Milne, Gordon. *George William Curtis & the Genteel Tradition*. Bloomington: Indiana University Press, 1956.

Mitchell, Catherine C., ed. *Margaret Fuller's New York Journalism: A Biographical Essay and Key Writings*. Knoxville: University of Kentucky Press, 1995.

Mitchell, Lee Clark. *Witness to a Vanishing America: The Nineteenth-Century Response*. Princeton, NJ: Princeton University Press, 1981.

Moldenhauer, Joseph J. "Editorial Appendix." In *Cape Cod*, edited by Henry David Thoreau and Joseph J. Moldenhauer, 237–441. Princeton, NJ: Princeton University Press, 2004.

Moldenhauer, Joseph J. "Editorial Appendix." In *Excursions*, edited by Henry David Thoreau and Joseph J. Moldenhauer, 291–650. Princeton, NJ: Princeton University Press, 2007.

Moldenhauer, Joseph J. "Editorial Appendix." In *The Maine Woods*, edited by Henry David Thoreau and Joseph J. Moldenhauer, 349–486. Princeton, NJ: Princeton University Press, 1972.

Morton, Timothy. *Ecology Without Nature: Rethinking Environmental Aesthetics*. Cambridge, MA: Harvard University Press, 2007.

Moskowitz, Alex. "Economic Imperception; or, Reading Capital on the Beach with Thoreau." *American Literary History* 32, no. 2 (2020): 221–42.

Murray, Charles. *Travels in North America During the Years 1834, 1835, & 1836 Including a Summer Residence with the Pawnee Tribe of Indians, in the Remote Prairies of the Missouri, and a Visit to Cuba and the Azore Islands*. London: R. Bentley, 1839.

Murray, William Henry Harrison. *Adventures in the Wilderness, or, Camp-Life in the Adirondacks*. Boston: Fields, Osgood, 1869.

Myerson, Joel, ed. *Critical Essays on Margaret Fuller*. Boston: G.K. Hall, 1980.

Myerson, Joel. *Margaret Fuller: A Descriptive Bibliography*. Pittsburgh, PA: University of Pittsburgh Press, 1978.

Myerson, Joel. "Mrs. Dall Edits Miss Fuller: The Story of Margaret and Her Friends." *Papers of the Bibliographic Society of America* 72, no. 2 (1978): 187–200.

Myerson, Joel. "Thoreau, Hawthorne, and the 'Seven-Mile Panorama.'" *ESQ: A Journal of the American Renaissance* 44, no. 4 (1998): 226–73.

Nash, Roderick. *Wilderness and the American Mind*. New Haven, CT: Yale University Press, 1967.

Newman, Lance. *Our Common Dwelling: Henry Thoreau, Transcendentalism, and the Class Politics of Nature*. New York: Palgrave Macmillan, 2008.

Newman, Lance. *The Literary Heritage of the Environmental Justice Movement: Landscapes of Revolution in Transatlantic Romanticism*. New York: Palgrave Macmillan, 2019.

Nicolar, Joseph. *The Life and Traditions of the Red Man*. Edited by Annette Kolodny. Durham, NC: Duke University Press, 2007.

Noe, Marcia. "The Heathen Priestess of the Prairie: Margaret Fuller Constructs the Midwest." *Old Northwest: A Journal of Regional Life and Letters* 16, no. 1 (Spring 1992): 3–12.

Novack, Barbara. *Nature and Culture: American Landscape and Painting, 1825–1875*. Oxford: Oxford University Press, 1980.

O'Connell, Barry, ed. *On Our Own Ground: The Complete Writings of William Apess, A Pequot*. Amherst: University of Massachusetts Press, 1992.

"Ojibwe Treaty Rights." Wisconsin Indian Research Project. Milwaukee, WI: Milwaukee Public Museum. www.mpm.edu/content/wirp/ICW-110. Accessed April 7, 2022.

Peacock, Thomas D., Marlene Wisuri, and Winona LaDuke. *Ojibwe Waasa Inaabidaa: We Look in All Directions*. Afton, MN: Afton Historical Society Press, 2002.

Pearce, Roy Harvey. *Savages and Civilization: A Study of the History of the Indian and the American Mind*. Berkeley: University of California Press, 1988.

Peyer, Bernd C. *The Tutor'd Mind: Indian Missionary-Writers in Antebellum America*. Amherst: University of Massachusetts Press, 1997.

Peyser, Thomas. "*Walden*'s Sandbank and the *Vestiges* Controversy." *American Notes and Queries* 30, no. 1 (2017): 126–36.

Pickford, Benjamin. "*Cape Cod*, Literature, and the Illocality of Thinking About Capital." In *Thoreau Beyond Borders: New Essays on an International Essays on America's Most Famous Nature Writer*, edited by François Specq, Laura Dassow Walls, and Julien Nègre, 179–94. Amherst: University of Massachusetts Press, 2020.

Pinsky, Robert. "Comedy, Cruelty, and Tourism: Thoreau's *Cape Cod*." *American Scholar* 79, no. 3 (2004): 79–88.

Poe, Edgar Allen. *Essays and Reviews*. New York: Library of America, 1984.

Pourtalès, Albert-Alexandre. *On the Western Tour with Washington Irving: The Journals and Letters of Count de Pourtalès*. Edited by George F. Spaulding. Norman: University of Oklahoma Press, 1968.

Randolph, Richard S. "'I Am of French Extract': Thoreau's Sympathy with the French." *Concord Saunterer* 6 (1998): 46–62.

Rex, Cathy. "Survivance and Fluidity: George Copway's *The Life, History, and Travels* of Kah-ge-ga-gah-bowh." *Studies in American Indian Literatures* 18, no. 2 (Summer 2006): 1–33.

Richardson, Robert D., Jr. *Henry Thoreau: A Life of the Mind*. Berkeley: University of California Press, 1986.

Rigby, Kate. *Dancing with Disaster: Environmental Histories, Narratives, and Ethics for Perilous Times*. Charlottesville: University of Virginia Press, 2015.

Robinson, David. *Natural Life: Thoreau's Worldly Transcendentalism*. Ithaca, NY: Cornell University Press, 2004.

Robinson, David. "Thoreau's 'Ktaadn' and the Quest for Experience." In *Emersonian Circles*, edited by Robert Burkholder and Wesley T. Mott, 207–24. Rochester, NY: University of Rochester Press, 1997.

Rose, Suzanne Dvorak. "Tracking the Moccasin Print: A Descriptive Index to Henry David Thoreau's Indian Notebooks and a Study of the Relationship of the Indian Notebooks to Mythmaking in Walden." PhD diss., University of Oklahoma, 1994.

Rosowski, Susan J. "Margaret Fuller, an Engendered West, and *Summer on the Lakes*." *Western American Literature* 25, no. 2 (Summer 1990): 125–44.

Rossi, William. "Making *Walden* and Its Sandbank." *Concord Saunterer* 30 (2022): 10–58.

Royle, Nicholas. *Veering: A Theory of Literature*. Edinburgh: Edinburgh University Press, 2011.

Ruoff, A. LaVonne Brown. "The Literary and Methodist Contexts of George Copway's *Life, Letters, and Speeches*." In *Life, Letters, and Speeches*, edited by Donald B. Smith and A. LaVonne Brown Ruoff, 1–21. Lincoln: University of Nebraska Press, 1997.

Sattelmeyer, Robert. *Thoreau's Reading: A Study in Intellectual History with Bibliographical Catalogue*. Princeton, NJ: Princeton University Press, 1988.

Sayre, Robert F. *Thoreau and the American Indians*. Princeton, NJ: Princeton University Press, 1977.

Schlett, James. *A Not Too Greatly Changed Eden: The Story of the Philosophers' Camp in the Adirondacks*. Ithaca, NY: Cornell University Press, 2015.

Schmalz, Peter S. *The Ojibwa of Southern Ontario*. Toronto: University of Toronto Press, 1991.

Schneider, Richard. *Civilizing Thoreau: Human Ecology and the Emerging Social Sciences in the Major Works*. Rochester, NY: Camden House, 2016.

Schneider, Ryan. "Drowning the Irish: National Borders and Class Boundaries in Henry David Thoreau's *Cape Cod*." *American Transcendental Quarterly* 22, no. 3 (2008): 463–76.

Schoolcraft, Henry Rowe. *Algic Researches: Comprising Inquiries Respecting the Mental Characteristics of North American Indians*. 2 vols. New York: Harper & Bros, 1839.

Schulz, Dieter. "Nature, Knowledge, and the Method of Thoreau's Excursions." In *Thoreauvian Modernities: Transatlantic Conversations on an American Icon*, edited by Francois Specq, Laura Dassow Walls, and Michel Granger, 173–86. Athens: University of Georgia Press, 2013.

Sears, John F. *Sacred Places: American Tourist Attractions in the Nineteenth Century*. New York: Oxford University Press, 1989.

Sexton, Melissa. "Sublime Matter: Materiality and Language in Thoreau's Ktaadn." In *Rediscovering the Maine Woods: Thoreau's Legacy in an Unsettled Land*, edited by John J. Kucich, 105–22. Amherst: University of Massachusetts Press, 2019.

Shanley, J. Lyndon. *The Making of Walden, with the Text of the First Edition*. Chicago: University of Chicago Press, 1957.

Simon, Katie. "Affect and Cruelty in the Atlantic System: The Hauntological Argument of Henry David Thoreau's *Cape Cod*." *ESQ: A Journal of Nineteenth Century Literature and Culture* 62, no. 2 (2016): 245–82.

Smith, Donald B. "Kahgegagahbowh: Canada's First Literary Celebrity in the United States." In *Life, Letters, and Speeches*, edited by George Copway, Donald B. Smith, and A. LaVonne Brown Ruoff, 23–60. Lincoln: University of Nebraska Press, 1997.

Smith, Donald B. "The Life of George Copway." *Journal of Canadian Studies* 23, no. 3 (Fall 1988): 5–38.

Smith, Donald B. *Mississauga Portraits: Ojibwe Voices from Nineteenth-Century Canada*. Toronto: University of Toronto Press, 2013.

Smith, Susan Belasco. "Introduction." In *Summer on the Lakes*, edited by Margaret Fuller and Susan Belasco Smith, vii–xxii. Urbana: University of Illinois Press, 1991.

Speck, Frank G. "The Eastern Algonkian Wabanaki Confederacy." *American Anthropologist* 17 (1915): 492–508

Speck, Frank G. *Penobscot Man: The Life History of a Forest Tribe in Maine*. Orono: University of Maine Press, 1997. Originally published 1940.

Stoneham, Michael. "Remeasuring Thoreau: *The Maine Woods* and Thoreau's Evolving Appreciation of the Racial Other." *Concord Saunterer* 27 (2019): 68–88.

Stoneham, Michael. "Thoreau's Hunting Ethos: Teaching Right Natural Relations in Modern America." *Thoreau Society Bulletin* 282 (2013): 8–10.

Stowe, William W. "Conventions and Voices in Margaret Fuller's Travel Writing." *American Literature* 63, no. 2 (June 1991): 242–62.

Taylor, Bayard. *Views Afoot, or Europe Seen with Knapsack and Staff*. New York: Wiley and Putnam, 1846.

Taylor, Bayard. *Ximena, or the Battle of the Sierra Morena, and other Poems*. Philadelphia, PA: Herman Hooker, 1844.

Thoreau, Henry David. *Cape Cod*. Edited by Joseph J. Moldenhauer. Princeton, NJ: Princeton University Press, 1988.

Thoreau, Henry David. *The Correspondence of Henry D. Thoreau*. 2 vols. Edited by Robert N. Hudspeth, Elizabeth Hall Witherell, and Lihong Xie. Princeton, NJ: Princeton University Press, 2013.

Thoreau, Henry David. *Early Essays and Miscellanies*. Edited by Joseph J. Moldenhauer. Princeton, NJ: Princeton University Press, 1975.

Thoreau, Henry David. *Excursions*. Edited by Joseph J. Moldenhauer. Princeton, NJ: Princeton University Press, 2007.

Thoreau, Henry David. *Journal*. Edited by Elizabeth Hall Witherell, William L. Howarth, Robert Sattelmeyer, Thomas Blanding, Mark R. Patterson, William Rossi, Leonard N. Neufeldt, Nancy Craig Simmons, Patrick F. O'Connell, Heather Kirk Thomas, Ron Thomas, and Sandra Harbert Petrulionis, 8 vols. Princeton, NJ: Princeton University Press, 1981–.

Thoreau, Henry David. Journal Transcripts. Unpublished. 23:322 cf. http://thoreau.library.ucsb.edu/writings_journals23.html.

Thoreau, Henry David. *Letters to a Spiritual Seeker*. Edited by Bradley P. Dean. New York: Norton, 2005.

Thoreau, Henry David. *The Maine Woods*. Edited by Joseph J. Moldenhauer. Princeton, NJ: Princeton University Press, 1972.

Thoreau, Henry David. *Walden*. Edited by J. Lyndon Shanley. Princeton, NJ: Princeton University Press, 1971.

Thoreau, Henry David. *Walden: A Fluid Text Edition*. Digital Thoreau. https://digitalthoreau.org/fluid-text-toc. Accessed May 15, 2025.

Thoreau, Henry David. *A Week on the Concord and Merrimack Rivers*. Editd by Carl F. Hovde, William L. Howarth, and Elizabeth Witherell. Princeton, NJ: Princeton University Press, 1980.

Thorson, Robert. *The Boatman: Henry David Thoreau's River Years*. Cambridge, MA: Harvard University Press, 2019.

Thorson, Robert. *Walden's Shore: Henry David Thoreau and Nineteenth-Century Science*. Cambridge, MA: Harvard University Press, 2014.

Thwaites, Reuben Gold, ed. *The Jesuit Relations and Allied Documents: Travels and Explorations of the Jesuit Missionaries in New France, 1610–1791; the Original French, Latin, and Italian Text, with English Translations and Notes*, 73 vols. Cleveland, OH: Burrows Bros., 1896.

Tonkovich, Nicole. "Traveling in the West, Writing in the Library: Margaret Fuller's *Summer on the Lakes*." *Legacy: A Journal of American Women Writers* 10, no. 2 (1993): 79–102.

Trafton, Melissa Geisler. "'It Is a Joint Venture': John Frederick Kensett's Images for *Lotus-Eating*." *American Art* 25, no. 2 (2011): 104–19.

Tuan, Yi Fu. *Space and Place: The Perspective of Experience*. Minneapolis: University of Minnesota Press, 1977.

Vanden Bossche, Chris R. "Notes on the Text." *Past and Present* (2005): lxxi–ciii.

Vilhauer, Benjamin. "The Theme of Time in Thoreau's *Cape Cod*: The Crisis of the Present Shore." *Concord Saunterer* 16 (2008): 33–44.

Vizenor, Gerald. *Manifest Manners: Narratives on Postindian Survivance*. Lincoln: University of Nebraska Press, 1999.

Vogelius, Christa Holm. "*Cape Cod*'s Transnational Bodies." In *Thoreau Beyond Borders: New Essays on an International Essays on America's Most Famous Nature Writer*, edited by François Specq, Laura Dassow Walls, and Julien Nègre, 25–39. Amherst: University of Massachusetts Press, 2020.

Walls, Laura Dassow. *Henry David Thoreau: A Life*. Chicago: University of Chicago Press, 2017.

Walls, Laura Dassow. *Seeing New Worlds: Henry David Thoreau and Nineteenth-Century Science*. Madison: University of Wisconsin Press, 1995.

Walls, Laura Dassow. "'Some Star's Surface': Thoreau in the Maine Woods." In *Rediscovering the Maine Woods*, edited by John J. Kucich, 87–104. Amherst: University of Massachusetts Press, 2019.

Walls, Laura Dassow. "Walking West, Gazing East: Planetarity on the Shores of Cape Cod." In *Thoreauvian Modernities: Transatlantic Conversations on an American Icon*, edited by François Specq, Laura Dassow Walls, and Michel Granger, 21–42. Athens: University of Georgia Press, 2013.

Warren, James Perrin. *Thoreau's Botany: Thinking and Writing with Plants*. Charlottesville: University of Virginia Press, 2023.

Warren, William W. *History of the Ojibway People*. St. Paul: Minnesota Historical Society Press, 1984.

Warrior, Robert Allen. *Tribal Secrets: Recovering American Indian Intellectual Traditions*. Minneapolis: University of Minnesota Press, 1995.

Weaver, James. "The Continent Behind: Alienation and the American Scene in George William Curtis's *Lotus-Eating: A Summer Book*." *Journeys* 17, no. 1 (2016): 27–50.

Weisenberg, Michael C. "Beyond the Borders of Time: Thoreau and the 'Ante-Pilgrim History' of the New World." In *Thoreau Beyond Borders: New Essays on an*

International Essays on America's Most Famous Nature Writer, edited by François Specq, Laura Dassow Walls, and Julien Nègre, 40–54. Amherst: University of Massachusetts Press, 2020.

Williams, Jericho. "Trigger Warning: Thoreau, 'Chesuncook,' and the Complications of Hunting." *Thoreau Society Bulletin*, no. 286 (2014): 1–3.

Williams, Stanley T., and Barbara D. Simison. Introduction to *Washington Irving on the Prairie; or, A Narrative of the Tour of the Southwest in the Year 1832*. Edited by Henry Leavitt Ellsworth, Stanley T. Williams, and Barbara D. Simison, vii–xvii. New York: American Book Company, 1937.

Willson, Lawrence. "Thoreau: Student of Anthropology." *American Anthropologist* 61, no. 2 (April 1959): 279–89.

Witgen, Michael. *An Infinity of Nation: How the Native New World Shaped Early North America*. Philadelphia: University of Pennsylvania Press, 2012.

Womack, Craig S. *Red on Red: Native American Literary Separatism*. Minneapolis: University of Minnesota Press, 1999.

Wong, Hertha D. Sweet. "Native American Life Writing." In *Cambridge Companion to Native American Literature*, edited by Joy Porter and Kenneth M. Roemer, 125–44. Cambridge: Cambridge University Press, 2005.

Yusoff, Katherine. *A Billion Black Anthropocenes or None*. Minneapolis: University of Minnesota Press, 2018.

Zwarg, Christina. *Feminist Conversations: Fuller, Emerson, and the Play of Reading*. Ithaca, NY: Cornell University Press, 1995.

Index

www.ingramcontent.com/pod-product-compliance
Ingram Content Group UK Ltd.
Pitfield, Milton Keynes, MK11 3LW, UK
UKHW021850130926
475466UK00003B/48

9 798765 163566